2.15.77

BOMB SECURITY GUIDE

By Graham Knowles

Security World Publishing Co., Inc.
Los Angeles, CA 90034

Copyright © 1976 by Security World Publishing Co., Inc.

All rights reserved. No part of this book may be used or reproduced in any manner without written permission except in the case of brief quotations embodied in critical articles and reviews.

First Edition 1976

Security World Publishing Co., Inc.
2639 South La Cienega Boulevard
Los Angeles, California 90034

Printed in the United States of America

Library of Congress Cataloging in Publication Data

Knowles, Graham, 1949-
  Bomb security guide.

  Includes index.
  1. Bombs--Safety measures. 2. Industry--Security measures. I. Title.
HV8290.K56   658.4'7   76-41301
ISBN 0-913708-25-9

CONTENTS

Preface

Chapter 1   THE BOMB RISK 1
Recent Bomb Statistics • Bomb Threat Targets • Motivations of the Bomber • Organized Terrorism • The Terrorist Bomber • Roles of Public and Private Security • Target Risk Evaluation • Forms of Sabotage • Effects of Bomb Threats • Risks of Explosion • Need for Preventive Security Program

1956074

Chapter 2   PREVENTIVE SECURITY 15
Perimeter Security • Outer Barriers • Lighting and Patrol • Good "Housekeeping" • Windows and Other Openings • Closed Circuit Television • Entry Control • Staff Entry Control • Visitor Entrance Supervision • Checking Deliveries • Vehicle Control and Parking • Public Relations and Staff Acceptability in Searching • Internal Security Systems • Sealing Off Openings • Checking Vulnerable Areas • Visibility Deters Bomb Planter • Vital Area Protection • External Threats • Upgraded Security • Staff Deployment and Education in Bomb Security • Staff Training • A Cooperative Effort

Chapter 3   HAZARDOUS DEVICES 33
The Firing Train • Explosive Charges • Low Explosives • High Explosives • Incendiaries • Activating Explosive Devices • Time Delay Activations • Anti-Personnel Handling Activations • Blasting Caps: Electrical and Pyrotechnic • Safety Precautions

Chapter 4   TELEPHONED BOMB THREATS 47
Reported Bomb Damage • Telephone Bomb Threat Procedures • Operator Training in Bomb Threat Procedures • Remain Calm • Follow Pre-Planned Procedure • Get Vital Facts • Encourage Caller to Talk • Bomb Threat Report • Emergency Contact List

Chapter 5 BOMB THREAT SECURITY RESPONSE 55
Alternative Security Responses • Evacuation: Full or Localized • Clearing Evacuation Routes and Assembly Areas • Evacuation Instructions • Discreet Warnings • Use of Fire Alarms • Recommended Evacuation Distances • Bomb Threat Evacuation Methods and Instructions • Search Operations: By Employees, Security Personnel, Police • Search Safety Rules • Search Techniques • Clues to Recognition • Minimizing Explosion Damage • Vehicle Searching • Summary: Step-by-Step Emergency Response Procedure

Chapter 6 MANAGEMENT BOMB THREAT
PROCEDURES 81
Establishing Policy • Basic Decisions • Reliance on Government Agencies • Staff Deployment to Preventive and Emergency Security Duties • Preventive Security • Emergency Responses • Emergency Officer

Chapter 7 LETTER BOMB SECURITY 89
Target Risk Evaluation • Developing A Security System • Staff Education • Detection Devices: Metal Detectors • X-Ray Examiners • Explosive Vapor Detectors • Recognition of Hazardous Mail Devices • Detection of Letter Bombs • Recognition Clues • Security Response to Suspect Hazardous Mail Devices: If Device Has Not Been Touched • If Item Is Suspected During Handling • Letter and Parcel Bomb Guide

Appendix A BOOBY TRAPS, MINES & IMPROVISED
EXPLOSIVE DEVICES 103

Appendix B DRAMATIZED BOMB THREAT SECURITY
RESPONSE 117

Appendix C OPERATION OF X-RAY INSPECTION
EQUIPMENT 139

Appendix D BOMB THREATS — POLICE
RESPONSE PROCEDURE 141

INDEX 153

PREFACE

This bomb threat emergency response guide has been developed primarily to assist organizations and their security officers in contingency planning for bomb threats and incidents.

The first section introduces the bomb risk situation as it exists in society today, and provides essential data on perpetrators of threat calls and their motivations. It is followed by a comprehensive guide to preventive security systems, the essential basis of an anti-bomb threat program.

Succeeding chapters cover the recognition of hazardous devices and steps to minimize device activation risks; telephone operators' reporting procedures; a basic security response plan, including the key element of evacuation methods and evaluation and search procedures; and the special problems of letter bombs. A step-by-step emergency response check list is incorporated, which may be used for tactical operation exercises.

A bomb threat security response requires cooperation between internal security and law enforcement agencies, fire and medical departments, and sometimes even military explosive ordnance personnel (EOD) from the armed services. An effective bomb threat response program requires integrated planning and inter/intra-departmental cooperation. For this reason it is essential that the security plan receive the full support of an organization's administration.

A preventive security system may be set up by an organization's own security director or by an outside security consultant. In either case, in this writer's view it is beneficial to have an external consultant evaluate an organization's security effectiveness. When developing preventive security programs, one should think of the tactics from the terrorist's or bomber's point of view. How would you penetrate the organization's perimeter and plant a device? Such questions are often most efficiently answered by an independent outside consultant, who is also in a position to make objective recommendations (as in the expensive area of hardware) for developing a cost-effective security program.

While this manual provides essential planning and operational information, it cannot provide efficiency in its use. Each organization must effect an ongoing program of training and evaluating its security efficiency. There is no substitute for operational exercises and staff education. Training and preparation are key elements in an effective anti-bomb threat program.

The major objective of telephoned bomb threats is usually psychological and financial harassment. Time spent in testing and practicing security responses will be amply returned in time saved during actual bomb threat situations.

Finally, the liability that might be held to exist for damages and personal injury resulting from a bomb incident will vary in different areas and specific circumstances. Organizations are therefore urged to investigate that potential liability in the applicable jurisdiction. But whatever the legal and insurance considerations that apply, the general responsibility of a company to provide for the safety and well-being of its employees and others on its premises offers another compelling argument, if one were needed, for developing and instituting a sound bomb security policy and procedure, and for thorough training of responsible personnel.

Graham Knowles
Cambridge, England

Chapter 1
THE BOMB RISK

Excerpt from FBI Uniform Crime Reports:

BOMB HIGHLIGHTS — 1975

A total of 2,074 bombing incidents were reported to the FBI for the year 1975. In 1974, a total of 2,044 incidents occurred.

Two thousand four hundred and nineteen devices were used in connection with the 2,074 bombing incidents. Sixty percent or 1,451 were explosive in nature, while 40 percent or 968 were incendiary.

Sixty-nine deaths and 326 injuries were reported to have occurred as a result of bombing incidents. Two hundred and seventy-three injuries and 45 deaths occurred in connection with explosive incidents, and 53 injuries and 24 deaths with incendiary.

The total value of property damaged due to bombings was $27,003,981. Explosive bombs resulted in $24,896,292 damage while incendiary devices caused $2,107,689 damage. *

In March, 1970, coordinated explosions extensively damaged the Manhattan skyscraper offices of Socony Mobile Oil, General Telephone & Electronics, and International Business Machines (IBM). These attacks seemed to signal a new phase of urban guerrilla warfare and terrorism to usher in the decade of the 1970's. The weapons chosen for terrorist actions, primarily by left-wing organizations, included assassination, kidnaping and psychological warfare. Psychological and actual sabotage were achieved by means of explosive

*FBI Uniform Crime Reports, *Bomb Summary: A Comprehensive Report of Incidents Involving Explosive and Incendiary Devices in the Nation—1975* (United States Department of Justice, 1976), p. 3.

bombs, incendiary devices and telephoned bomb threats. One of the most dramatic incidents involved a bomb which damaged the U.S. Capitol building.

In the years that followed the 1970 wave of bombings, such explosive attacks have occurred on an almost daily basis. The escalating situation was the subject of a Senate Subcommittee inquiry chaired by Senator John L. McClellan. For a 105-day period from January to April, 1972, the committee received reports on 4,330 bombings and arson attacks, plus 1,475 attempted bombings and 35,129 bomb threats. Property damage was estimated at a minimum of $21 million. Despite the fact that militants claimed attacks were made only against property, during the period in question 40 people were killed and 384 injured and maimed by explosive attacks.

During 1972 the New York Police Department received over 10,000 bomb reports. This was three times as many as in 1970, and at least ten times as many as in 1968, just four years earlier. Only three percent of the calls involved actual explosives, and the total of malicious false alarms was approximately 9,700. For the 105-day period studied by the Senate Subcommittee, the man-hour loss figure for New York City alone from malicious false alarm bomb threats was estimated at $350 million ... a figure which, as events have proved, the beleaguered city could hardly afford.

This steadily rising pattern of bombing incidents has continued, as illustrated in Figure 1. The unusually large number of deaths and injuries, and the high damages to property, in 1975 were the result of three major incidents among the 2,074 reported for the year. One was a January, 1975, explosion in a commercial building in New York City. Another involved the detonation of three explosive devices in an industrial building in Connecticut, and the third was the December explosion at LaGuardia Airport in New York. In these three incidents alone 15 persons died, 107 more were injured, and property damage was in excess of $15 million.

The siege continues. Daily the news media demonstrate that the threat of explosive or incendiary attack against commercial, industrial, government and political organizations is real. In the continuing escalation of urban guerrilla/terrorist operations in countries throughout the world, almost any organization or event can become the target of a bomb or bomb threat. No organization can afford to

Year	Total Actual and Attempted Bombings	Actual		Attempt		Property Damage (Dollar Value)	Personal Injury	Death
		Explo.	Incend.	Explo.	Incend.			
1972	1,962	714	793	237	218	7,991,815	176	25
1973	1,955	742	787	253	173	7,261,832	187	22
1974	2,044	893	758	236	157	9,886,563	207	24
1975	2,074	1,088	613	238	135	27,003,981	326	69

Figure 1. Bombing Incidents: 1972 through 1975. Data from FBI Uniform Crime Reports, *Bomb Summary: A Comprehensive Report of Incidents Involving Explosive and Incendiary Devices in the Nation—1975,* United States Department of Justice, 1976.

be complacent about the risks of a bomb threat to their personnel or resources. No organization is immune.

Bomb Threat Targets

Bomb attacks, threats and hoaxes are received by offices, factories, hospitals, transport stations and terminals, libraries, exhibitions and entertainment centers. A bomber may strike against an obvious target, such as a government munitions factory, where he will achieve politico-military advantages; or against an old people's home or children's party, where the advantage to be gained is purely psychological. He may even strike for no visible motive, as in the case of the bomb which exploded in the baggage claim area at LaGuardia Airport on December 29, 1975, killing at least 11 persons and injuring about 75 others.

Although any organization may receive a bomb threat, there are some more likely and therefore high-risk terrorist targets. Examples of these are government and political centers, buildings containing civil dignitaries or VIP's, defense-related industries, fuel and chemical installations, power stations and the like. To this civil list must be added almost any military target, from an active installation to vehicles in transit to recruitment offices and ex-servicemen's clubs.

The primarily left-wing underground press frequently provides explicit details of various factories and organizations which they consider to be unacceptable parts of the free world "system." By these and other subversive methods, potential targets are indicated to terrorist cells and lone extremists.

High-risk targets must, of course, be especially prepared and their staffs trained to counter the risks of bombing. A high-risk target may have to integrate its security response plan with local disaster forces, as in the case of fuel or chemical installations. While the material damage from bombings is usually minor, a strike against a high-risk installation could cause hundreds of injuries and widespread physical destruction.

Motivations of the Bomber

A bomber or arsonist may be motivated by many different factors. These include racial strife, color, a specific terrorist campaign, religion, personal animosity, crime, political and quasi political strife, mental illness, and many other real or imagined sources of grievance.

Because of these widely differing motivations, a bomber may vary considerably in his intellectual level and psychological profile. His abilities as a terrorist will also be variable. Hazardous devices encountered may vary from a simple, improvised device to a complex, booby-trapped bomb. Information and instructions on the use of explosives are available through courses in Communist countries, subversive and radical groups organized within our own country, and the underground press.

Bomb hoaxes usually have clearly defined motivations. Hoax calls are most commonly received from dissatisfied employees, overt political groups and extremists, unthinking or drunken practical jokers, mentally disturbed individuals—and even children.

Unfortunately, it is not often possible to determine whether a threat call originates from a harmless crank or a determined terrorist. Each threat, therefore, should be considered as a genuine danger until the fact that it is a hoax can be established.

Organized Terrorism

As indicated above, bombings are perpetrated by terrorists, criminals, mentally deranged persons, or even those acting out of

Target	No. of Incidents	Actual Explo.	Actual Incend.	Attempted Explo.	Attempted Incend.
Residences	582	234	255	42	51
Commercial Operations	485	275	127	56	27
Vehicles	273	134	70	47	22
School Facilities	165	87	48	18	12
Law Enforcement	76	31	27	12	6
Government Property	62	37	13	9	3
Persons	43	26	4	12	1
Public Utilities	41	33	1	6	1
Recreation Facilities	33	21	4	4	4
Communication Facilities	32	31	1	–	–
Other	282	179	63	32	8
TOTALS	2074	1088	613	238	135

Figure 2. Bombing Incidents by Target – 1975. Data from FBI Uniform Crime Reports, *Bomb Summary*, **pp. 5-6.**

Apparent Motive	No. of Incidents	Casualties Injury	Casualties Death
Malicious Destruction	745	38	–
Personal Animosity	723	93	38
Unknown Motive	192	74	13
Labor Dispute	75	2	1
Extremist	73	7	–
Political (U.S./Foreign)	62	66	6
Monetary Gain	57	9	4
Civil Rights	47	–	–
Anti-Establishment	46	15	1
Racketeering	16	1	2
Other	38	21	4
TOTALS	2074	326	69

Figure 3. Bombing Incidents by Apparent Motive – 1975. Data from FBI Uniform Crime Reports, *Bomb Summary*, p. 16.

personal animosity. Terrorist actions, however, account for most real bombing incidents. It may therefore be useful to examine briefly the makeup and purposes of organized terrorist groups.

Terrorism may be defined as "violent criminal activity, designed to intimidate for political purposes." Terrorist actions are intended to focus attention on a particular cause. Bombings and bomb threats are used to de-stabilize economy, to reduce the public's confidence in the government or police, to cause terror and thus dramatize a grievance.

Theories of revolutionary warfare show that there are at least five phases of a complete revolutionary campaign. Terrorism is normally the third phase.

An organizational phase comes first, in which unions and societies are formed, infiltrated into sectors of industry, students and public life. These are gradually prepared and motivated toward forms of revolution. Next comes a phase of political action, in which the masses or minority groups are motivated and approached for support. This phase includes financial de-stabilizing tactics such as strikes, work stoppages and other forms of indirect sabotage.

This is usually followed by a period of prolonged or intermittent terrorist activity. Such actions may include assassination, kidnaping, air or sea craft hijacking, bombing, arson or bomb threats. As attention or support is focused upon the revolutionary aims, the campaign may then escalate into guerrilla warfare.

This fourth phase, guerrilla warfare, may be divided into two forms, urban and rural warfare. These differ not only in locality but also in some tactics. Guerrilla tactics are generally more diversified than those used by terrorists; the guerrilla may use terrorism, political maneuvering, or many other means to achieve his ends. Bombing may be one of them, but it is mainly a terrorist act directed against government or its delineations.

A period of successful guerrilla warfare may be followed by a limited mobile warfare phase, fought by "conventional" weapons and tactics. However, at any time the revolutionaries may adopt one or more of the former phases to suit the tactical situation.

It should be emphasized that terrorism can quickly escalate into phases of guerrilla or even limited mobile warfare. The recent history of guerrilla actions, ranging from Indo-China to South America, Africa and some European countries, reveals that a major contributive factor in successful revolutionary campaigns was the ability of the revolutionaries to build up their organization, communications, expertise, weapons and manpower well before government or police forces realized the revolutionaries' strength and scale. Thus, when the political and tactical climate was right, the terrorist phase moved quickly into guerrilla warfare, catching government forces unprepared, ill-equipped and often undermanned.

The Terrorist Bomber

The very nature of revolutionary parties requires a secure structure of organization. Usually, terrorist units are designed for and capable of independent operation from the main party or organization. This ensures against discovery by government or police forces, and can also prevent publicly unacceptable acts of terrorism from being directly associated with the revolutionary party.

Terrorist units are usually commanded by highly trained and politically motivated and educated leaders, working from behind the front action line. The perpetrator of terrorist actions, however, is not often a skilled or highly educated guerrilla. The perpetrator's involvement with the terrorist unit may be politically inspired, but he may also have deeper motivations, such as an assumed glamorous or exciting group association. This is true especially if the unit has a reputation or public image, like the I.R.A. (Irish Republican Army), Bader-Mienhoff Gang, Angry Brigade, Symbionese Liberation Army, etc.

The bomber tends to be an individual with a weak personality, whose life has mainly been a failure, or at least non-distinctive. His ego may therefore be boosted by newspaper and television publicity. Bombing is a way of proving himself and establishing a viable self-image. It is for this reason that publicity about bomb incidents should be minimized wherever possible, and certainly not turned into media shows that glamorize group identities. Terrorism is, as we have defined it, merely violent criminal activity; it should not be allowed to develop martyrs or heroes.

Terrorist units frequently use the type of individual described to perpetrate their actions. Often the actual bomb planter or assassin has little real political connection with the revolutionary party. He is used and maneuvered, but he enjoys the glamorous associations and group identity, and may also feel that his actions are making a personal impact upon society.

The materials and training for bombing are easily obtained. Numerous subversive groups provide training in the construction and tactics of hazardous devices. Devices are obtained from stolen military or commercial sources, or by improvisation from common chemicals. The underground press, as we have indicated, not only frequently details targets in government and industry, but also disseminates detailed information on the construction and use of explosive devices. Further, international links exist between many subversive groups, for the communication of political information, explosives, weapons and instructors.

Roles of Public and Private Security

A bomb or bomb threat, particularly one which is part of a terrorist campaign, has two targets. The first, which may be a private organization's property or facility, is the physical location against which the attack or threat is made. The second target is society itself or the particular government. We have already made note of the economic de-stabilizing effects caused by bombings and bomb threats. It follows that the obviously damaging effects of such terrorist activities must be minimized by a combined public and private security counter-force.

To provide an effective counter-force, the separate roles of government police forces and private security should be established and understood, so that their roles may complement each other.

The main role of the police or military is intelligence acquisition, local counter-actions, and area security. For public police forces to concentrate upon these operations, they require coordinated assistance from private security agencies.

The role of internal security is essentially in the hands of private security. The greater the efficiency of private security in handling internal bomb security, the less time is taken away from public police for their maintenance of external security. Seen in this light, the development of internal bomb security systems and personnel training is not only essential for an organization's own protection, but it can also have a major effect upon external security affairs and the very stability of society.

Target Risk Evaluation

The primary stage in developing a security program of any kind is risk assessment. Is the organization a potential bomb target? Would bomb damage be unacceptable? What is the nature and extent of the potential risk? Is the organization in question a high-risk target, or one with unacceptably high personnel casualty probabilities, such as leisure, entertainment and public service industries?

Telephoned bomb threats and explosive incidents are best considered as acts of sabotage. There are six basic forms of sabotage:

1. *Mechanical:* breakage, the insertion of abrasives, inserting foreign bodies, failure to lubricate, maintain and repair, omission of parts.
2. *Chemical:* the insertion or addition of destructive, damaging or polluting chemicals in supplies, raw materials, equipment, product or utility systems.
3. *Fire:* ordinary means of arson, including the use of incendiary devices ignited by mechanical, electrical or electronic means.
4. *Electric/ Electronic:* interfering with or interrupting power, jamming communications, interfering with electric and electronic processes.
5. *Explosive:* damage or destruction by explosive devices; the detonation of explosive raw materials or supplies.
6. *Psychological:* riots, mob activity, the fomenting of strikes, jurisdictional disputes, boycotts, unrest, personal animosities, including excessive spoilage, doing inferior work, causing slowdown of operations, provocation of fear or work stoppage by false alarms, character assassination, bomb threats.

For the purposes of bomb security risk evaluation, we are concerned with three of the above areas of sabotage. These are (3) fire, (5) explosive, and (6) psychological sabotage.

What are the risks involved in a bomb threat or explosion? If we analyze these hazards, we can develop appropriate preventive security and emergency procedure systems to neutralize them.

Briefly, the potential effects of a telephoned bomb threat are:

1. A diversion for a crime
2. Panic
3. Loss of public and/or staff confidence
4. Loss of production time
5. Evacuation injuries through panic and confusion
6. Psychological harassment and stress

The risks of an explosion, in addition to all of those involved in a bomb threat listed above, can be summarized as follows:

1. Deaths
2. Severe injuries
3. Material damage
4. Cosmetic damage
5. Structural damage

Preventive security programs are designed to reduce these risks to acceptable levels. However, a manpower response is necessary for control and investigation of threat situations. Since the cost of security manpower is high, personnel numbers must be minimized, and those involved must be trained to control situations with such speed and efficiency as to reduce their damaging and de-stabilizing effects.

The second stage of risk evaluation, then, involves analyzing an organization's capability to handle bomb threat situations. The following questions should be considered. Do your security personnel:

1. Know how to prevent a devious bomber's entry?
2. Have explosive detection equipment?
3. Know how to handle bomb threat calls?
4. Know how to evaluate real and hoax threat calls?
5. Know how to conduct a safe evacuation?
6. Know the difference between fire and bomb alert procedures?

7. Know how to recognize explosives?
8. Know how to search for devices?
9. Know what actions to take if a bomb is reported?
10. Know how to deal with suspected vehicle devices?
11. Know how to cope with booby trap devices?
12. Know what triggers explosives?
13. Know how far personnel must be evacuated for certain sized bombs?
14. Know how to screen and handle explosive mail devices?

Security personnel training must cover all of these questions and many more if their response is to neutralize the substantial risks to personnel, property and economic stability posed by bomb threats.

Need for Preventive Security Program

As we have seen, the 1970's have witnessed a rapid escalation of bomb threats and actual bomb explosions, along with a world-wide increase in individual and organized terrorist activity. Terrorism has become a disease of society, often approaching epidemic proportions. And, like most other diseases, its best cure lies in prevention.

The vast majority of telephone bomb threats received are false, and except in the case of high-risk targets the probability of an actual explosion is very low. However, the substantial number and types of explosions which have occurred necessitate that every threat be treated as real until proven otherwise. Damage to property and time loss through bombings and bomb threats have already produced millions of dollars of loss each year, and all evidence would suggest that the future promises more of the same—and worse.

Many of the risks associated with bomb threats can be reduced with a sound preventive security system. Obviously, if we can prevent a bomber from entering an organization, we can eliminate most explosive risks along with the need for investigating telephoned bomb threats. If the security staff are reasonably sure that a bomber cannot have gained access to an area, because of the existence of an effective perimeter protection system, for example, then the process of dealing with a bomb threat is considerably simplified. There are also many quite minor security design concepts (to be outlined later in this book) which save considerable amounts of time during bomb threat situations.

The saving of lost time during bomb threats can alone make preventive security a cost-effective program. The difference between responding to a bomb threat when personnel have a practiced emergency response system, and handling a call unprepared, can be easily translated into "bottom line" savings in productive man-hours.

It should also be noted, recalling the brief psychological profile of terrorist bombers provided earlier, that the typical bomber is quite easily deterred by the presence of preventive security systems.

In summary, then, protection against explosive/incendiary damage and time loss can be achieved by evaluating the risks and developing an appropriate security program. Preventive and emergency response plans should be devised and *practiced*, so that personnel can react quickly to an emergency situation.

Not all of the security preventive and response procedures in the following pages will be necessary or possible in every organization, since not all facilities will be able, for example, to exercise the same degree of access control or personal search procedures. Here the particular organizational management or security director must be selective, and the risk assessment and the character of the facility will be the determining factors. However, the *principles* of bomb protection outlined here, ranging from the need for visibility to the planned telephone bomb threat response, will at least to some extent be applicable to all organizations.

Three catch phrases aptly sum up risk evaluation and contingency planning, and they should be remembered through each phase of the development of a bomb threat security program:

Complacency Causes Casualties
Planning Prevents Panic
and
Confusion Costs Money

Chapter 2
PREVENTIVE SECURITY

Preventive security has two separate functions against bomb risks. First, it reduces the risk both of external explosions and of penetration and bomb planting inside the organization. The secondary result of this is that, if a bomb threat call is received, the organization can be reasonably sure that it is a hoax, so that only a minimal response has to be organized. With only 2% of reported bomb threat calls being real, this makes preventive security a cost-effective program.

Preventive security falls into three main areas. These are Perimeter Protection, Internal Security, and Vital Area/Personnel Security. In this chapter we will consider each of these levels of protection in detail.

PERIMETER SECURITY

The prevention of unauthorized access is a major part of any security program, and especially bomb security. However, whereas the unauthorized removal of product is a risk with most security problems, here we are concerned with the unauthorized *entry* of hazardous devices.

An important aspect of perimeter security is its deterrent effect. The perpetrators of bombings tend to be easily deterred by obvious forms of physical and passive security. They prefer to attack unguarded and easily accessible installations. Therefore, the more obvious are the physical and passive security systems, the better their deterrent effect.

Installations obviously vary considerably in their layout and location. Not every installation is ideally designed or sited for high security. Securing against unauthorized entry on a new industrial site, or a remote military base, is far simpler than for an organization whose buildings are against a busy street. In the latter case, several perimeter protection systems may have to be used to achieve effective security.

Outer Barriers

Initial perimeter security is usually provided by a boundary wall or fence. As far as bomb security is concerned, open mesh fences are preferable to solid walls, as they allow easy vision by security guards. However, nearly all perimeter barriers are vulnerable to cutting or climbing, even when protected by barbed wire. They do provide a physical deterrent, because of the time involved in penetrating the barrier. Additional protection may be provided by utilizing motion detector alarms or complete electrical circuit alarms along the barrier. Where public access to the barrier is rare, alarms are a worthwhile investment.

The perimeter barrier and surrounding areas, both inside and outside of the installation, should be adequately illuminated at night. This provides a strong deterrent to unauthorized entry. Lighting is essential if internal security officers, and local police patrols, are to detect devices left at or near the perimeter.

Whatever the type of perimeter, it should be frequently patrolled by alert guards. This applies both to remote sites and to those where buildings adjoin public streets. The street should be frequently patrolled and a watch kept for abandoned vehicles or suspicious packages.

Good "Housekeeping"

To simplify patrol observation, and to avoid wasted time on active bomb searches, it is particularly important that both sides of the perimeter barrier be kept free of rubbish, trash containers and cartons, and concealing foliage or plantings. These all provide places

Commonplace office exterior offers the bomb planter several hiding places: inside hollow steel supports; on top of canopy; in flower bed; in trash container.

Ground floor of building is protected by blast curtains; perimeter is kept clear to facilitate anti-bomb patrols.

Two good bomb planting sites: trash can and sewer trap.

Untidy housekeeping leads to easy bomb planting. This area should be sealed off with wire mesh and kept locked.

for a bomber to plant a device. The elimination of litter around the perimeter barrier will cut active bomb search time from minutes to seconds.

If an organization's buildings have publicly accessible frontage, for example, a street or sidewalk, this too should be kept free of all objects. Such non-essential decorations as flower boxes, planters, trash cans and ornaments should be eliminated or sealed off to prevent their use for planting bombs or incendiaries. Basically, any area where a bomb can be placed and remain unseen should be removed or sealed off.

Windows and Other Openings

Anti-bomb security must go beyond the elimination of litter and places of concealment, to cover window ledges, drain pipes, drains, service inlets, air conditioning inlets, and the like. Strong steel mesh, locked into position, should be used to protect these vulnerable openings.

Utilizing ornamental iron grilles or plasticized windows will prevent devices from being thrown into a building. This is an important consideration when protecting key personnel or valuable resources. Window protection is particularly important when buildings frequently contain VIP's. The old-fashioned Molotov cocktail—which is essentially an improvised hand grenade—is still a favorite terrorist weapon.

It should be noted that it is not necessary to make a building look like a prison to achieve near maximum security effect, providing that modern security design and materials are utilized. The use of various grille designs, discreet armored glass and plastics will all provide unobtrusive security.

Closed Circuit Television

To minimize the extremely high cost of security guard patrols, and to provide a maximum of bomb security, closed circuit television systems may be installed to provide continual surveillance over wide areas.

Where a manned security console is available, closed circuit TV systems present highly cost-effective protection. Cameras may play a multiple role, because they can be used for anti-theft duties as well as bomb observation. An advantage of cameras is that they may be used to discreetly observe publicly accessible areas, such as exterior streets and walls.

With CCTV, and most other preventive security systems, high deterrent effect may be achieved by publicly displaying warning notices. Thus, warning notices should be displayed at frequent intervals along the perimeter, and security warning notices placed visibly at every entrance point.

Entry Control

Bombings and bomb threats are tactically more effective when made during an organization's working hours. Therefore the bomber must try to penetrate the organization to plant his device while the entry points are manned by security or other staff. He will probably try to penetrate by deception, confidently walking past guards or receptionists—possibly by using false identity papers.

The bomb planter may be a man or woman of any age and appearance. He could be dressed as a mechanic, carrying a tool box, or appear as a businessman with a neat attache case. He or she could be delivering ice cream or collecting for a well-known charity.

To prevent access of the bomber or his devices, staff and visitor entry must be checked and controlled. In many organizations the entry of goods, mail and parcels must also be monitored, to prevent the introduction of hazardous devices. The first and foremost line of defense, then, is at the perimeter entry point or points.

Dependent upon the type of organization, there are two basic types of entry control supervision. These are using security guard checks and receptionists. To achieve efficiency with either, cost-effective security monitoring systems should be considered. These will provide both greater security and less time wastage.

The security risks at entry control may be summarized as follows:

1. Devices being carried into the organization on a person.
2. Devices entering in hand luggage.
3. Devices entering in vehicles.
4. Devices entering as goods.
5. Devices entering as mail type deliveries.

These risks may be countered and neutralized by one or more of the following security procedures, depending on the organization and (in the case of body searches) the jurisdiction.

1. *Devices carried on the person:*
   - Hand searching persons entering on a regular or irregular basis.
   - Utilizing mobile metal detection equipment.
   - Passing all persons through an airport type metal detection gate.
   - Utilizing mobile gas chromotography detectors.
2. *Devices carried in hand luggage:*
   - Hand searching all hand luggage.
   - Utilizing mobile metal detection equipment (usually ineffective).
   - Utilizing mobile gas chromotography detectors.
   - Passing luggage through X-ray monitors.
3. *Devices entering in vehicles:*
   - Visual searching (requires mechanic's inspection to be safe).
   - Utilizing mobile gas chromotography detectors.
   - Utilizing trained vapor-detector dogs.
   - Developing a preventive parking system.
4. *Devices entering as goods:*
   - Hand searching (time consuming).
   - Utilizing mobile metal detectors.
   - Utilizing mobile or fixed gas chromotography detectors.
   - Utilizing fixed X-ray monitoring systems.
5. *Devices entering as mail deliveries:*
   - Initial visual searching (non-efficient).
   - Utilizing fixed gas chromotography detectors.
   - Utilizing fixed X-ray monitor systems.
   - Utilizing fixed or mobile metal detectors.

As suggested above, there are three basic entry situations: entry of an organization's staff, visitors and goods. Each will require an alert and consistent entry control procedure.

Staff Entry Control

Staff entry control and supervision may be achieved by using identity cards, sealed in plastic. These are available from a wide variety of commercial sources, or produced in minutes with mobile equipment. Color codes may be used to designate the type and area of entry for which the person is authorized. However, especially in high-risk areas, it must be remembered that security cards can be quite easily forged—and it has all too often been demonstrated that it is possible to substitute a photo of a gorilla on the card and walk past inattentive security guards.

For high-risk areas, there are a variety of automatic entry systems available. These include magnetic and electronic key card entry systems, coded push-button combination locks, palm print recognition, and others. Their use will be determined by the risk assessment.

Visitors

An organization's visitors may include company representatives, maintenance workers, public officials—or a bomber in any one of these disguises.

Visitor entrance must be controlled and supervised, but in a discreet and courteous fashion. The security officer or receptionist's check should be for the visitor's identification, the reason for the visit, and the person or department he wishes to visit.

The guard or receptionist should make a preliminary check as to the visitor's authorization to enter, whether his access allows the entry of deliveries, vehicles or other hand luggage, the extent of his entry authorization, etc. After this preliminary check at the entry point, the guard or receptionist should contact the visitor's internal sponsor for authentication. The visitor may then be issued a temporary pass or I.D. badge, which should preferably be worn externally. The guard or receptionist in some instances, especially in a large complex, arranges for the visitor to be escorted to and from his internal destination point.

All entries—visitors, staff and service people—should be recorded. A daily log should be maintained recording visitor names, their origin, I.D. check, reason for visit, time, internal supervisor or sponsor. Additionally it is essential that any hand luggage, samples or other objects carried into the organization by a visitor are logged IN AND OUT of the facility. This is to prevent an attache case or parcel bomb being left inside.

ENTRY CONTROL LOG _____   GUARD _____
GUARD POINT _____   DUTY TIME _____

Date	Entry Time	Name	Address	Visit Reason	I.D. Check	Sponsor	Luggage	Vehicle Time Out

Entry Log check by _____ Supervisor

Entry Incident Reports

Figure 4. Check entry control log

A sample check entry log, recommended for use by security guards or receptionists at entry points, is shown in Figure 4.

Deliveries

To prevent hazardous devices entering among deliveries, these should be checked and opened at their reception point, and the supervisor should inform the security staff that the delivery is both authorized and non-hazardous.

Deliveries to an organization are as diverse as a truck loaded with sand to parcels and liquids. All deliveries should be made under the supervision of security guards or another responsible person. Deliveries should be checked against invoices and orders, and no unauthorized deliveries accepted *under any condition.* Vehicles and delivery men should be subject to the same entry screening as other visitors—no matter if their truck has arrived at exactly 12:40 p.m. each day for the past two years.

Delivery vehicles entering may have to be searched, and standard weighbridge checking techniques used to prevent theft. Vehicles must be searched in three phases: the occupants, load and the vehicle itself. It is a good idea, when searching or checking a vehicle, for the security guard to ask the driver to accompany him as he checks the trunk or interior. In such circumstances the guard may be able to detect emotional stress in the driver.

While inside the organization, the vehicle may be escorted or observed by closed circuit television systems.

Parcels, mail and other small deliveries can be examined by a variety of explosive device detectors, including mobile or fixed X-ray systems, and vapor and metal detectors. A separate chapter is devoted to letter bomb security.

Vehicle Control and Parking

For an organization to protect its installation fully, a vehicle entry and parking control system must be utilized. The vehicle should not be overlooked; it represents a large explosive and incendiary hazard. Automobile devices have been used recently during terrorist campaigns for assassination and other criminal acts. Automobile devices may carry large amounts of explosive, and the gasoline tank provides a secondary explosive and incendiary risk. Pound for pound, gasoline can be as destructive as high explosive.

To counter automobile device risks, the best solution is a preventive security program. Automobile access must be restricted, and parking allowed only at a safe distance from vital areas. The greater the distance of vehicles from buildings and personnel, the greater the security.

However, anti-explosive security must be integrated with basic anti-theft security. Low-cost security systems such as entry and exit automatic cards, wire fence enclosures and interior lighting, closed circuit television observation or guard patrols may be utilized to prevent theft as well as to provide bomb security.

As with persons, there are three vehicle entry situations—staff entry, visitor entry and goods entry.

Staff entry may be controlled with cost-effective magnetic or punched card lock barriers, frequently used in private vehicle parking lots. This is far less costly than utilizing manned guard services.

Staff vehicles should be identified with distinctive badges, produced with tamper-proof plastic thermoseals. The security department should maintain records of personal vehicles authorized to enter the premises, the type of access, personal details, etc. This will aid swift investigation should a vehicle be suspected of carrying a device or require rapid owner identification.

Visitor vehicle parking should ideally be arranged in a separate controlled area. This should be as far away from the installation's vital points as possible. The registration or license numbers and other details should be written in the entry log by the guard or receptionist at the entry point.

As mentioned with delivery vehicles, if it is thought expedient to search a visitor's vehicle, the driver should accompany the searcher. By so doing the guard may be able to detect emotional stress in the driver, such as excessive sweating, skin coloration, rapid breathing or other signs of nervousness.

It is expedient to search vehicles which must pass by, or be parked by, vital areas. However, with the number of possible hiding places in a vehicle, it must be remembered that hand searching is time-consuming and not totally effective. Explosive vapor detectors provide the most effective search aid.

Public Relations and Staff Acceptability in Searching

Where expedient security measures begin to infringe upon the rights of an individual, care should be taken to see that the security measures adopted are acceptable to personnel or visitors. This is particularly true with regard to entry control and searching.

It should not be forgotten that a visitor's first and most vivid impression of an organization is usually based upon his initial reception. Therefore, politeness and courtesy are key elements in entry control duties for guards or receptionists. It is essential that security and other staff assigned to this task perform their duties in a considerate and friendly manner. Their function is to assist visitors and to protect their safety as well as that of the organization. These qualities and objectives should be emphasized during staff training.

Searching of visitors and staff may be a necessary routine security measure, or adopted on a random basis to provide preventive security. Whichever is the case, the security objective should be made clear to all personnel. Searching is far more acceptable when its objective and its necessity are understood by all.

A factor in the acceptability of searching is time loss by personnel. It was discovered in airport security programs that searching and screening were generally acceptable providing they did not unduly delay passengers. In this respect, the best way to achieve fast searching is by using fixed or mobile explosive/metal detectors. Hand searching is slow, creates problems with women, and is boring for security guards. Remember that anything that is boring or tedious for the performer almost inevitably leads to a relaxation of standards.

The legal ramifications of search and seizure should be discussed with corporate counsel and thoroughly understood by all security personnel. While private security officers are not normally under the limitations of law enforcement with respect to searches, nevertheless the application of legal restrictions concerning searches should be investigated for the particular jurisdiction.

INTERNAL SECURITY SYSTEMS

Initial perimeter defense and entry control reduce the need for an internal security program but do not, of course, eliminate it. As most bomb threats refer to internal locations, this is an important part of the security program.

The main form of internal security against explosive hazards is achieved by preventive security design. Although our recommendations for security design involve some construction and existing building modification expenses, security must be cost-evaluated against the time loss in bomb threats.

Internal security is achieved when the areas where a device may be planted are removed or sealed off. Elimination of possible device planting sites has two functions. First, it obviously decreases the vulnerability of the installation to a bomber. Secondly, it reduces the number of places and the time required to search if a bomb is reported. Although the latter factor, the search time saved after a bomb threat is received, may seem minor, in commercial organizations it can be a substantial budgetary consideration.

Openings

Basic essentials in internal preventive security are to seal off such areas as false ceilings, attics or lofts, elevator shafts, air conditioning and heating ducts, alcoves and other openings. All such areas should be sealed off with steel mesh grilles and securely locked.

Grilles are preferable to insecure boarding because they allow quick observation of the area. If open mesh grilles are not used, then the area or opening should be sealed off by building construction.

Vulnerable Areas

Internal fittings are frequently used to secrete an explosive device. This is particularly common in reception areas, where a bomb may be planted in an ash can, ornament, flower or plant display. These vulnerable bomb planting areas should either be removed or made easily visible to receptionists and security patrols. Other potential bomb hiding places are trash cans, waste containers, in and over light fittings, toilets and cisterns, flower pots or boxes, sand buckets and other fire fighting equipment.

Reception areas, corridors and other publicly accessible areas such as rest rooms, waiting rooms and lavatories should be subject to frequent security guard checks. These are all potential bomb sites,

Any accessible entrance should be sealed off with small diameter mesh.

An easy place to plant a device. Any accessible inlet/outlet should be sealed off with locked mesh gates.

Two common bomb planting sites: fire protection equipment and trash chute.

Roof light and elevator service shaft are potential bomb planting sites.

because they allow an intruder unobserved time to plant a device. He may also enter many of these areas before he is subjected to a search.

Other examples of danger spots are organization sports and recreation facilities, meeting halls, cafeterias and restaurants, etc. These areas should be protected by internal security design, as outlined above, and also by frequent security patrols. Certainly it should be made a security routine to check and "clear" sports complexes, recreation centers and meeting halls before people enter these places in any numbers.

Visibility

The main factor which creates a bomb planting hazard is lack of visibility of the bomb site or the bomb planter. Internal security is therefore automatically increased if all areas are made easily observable by security patrol or TV monitors. For example, locked non-security doors may be fitted with small observation windows. Another simple expedient to enhance internal security is to shorten curtains so that all window ledges are quickly visible.

As usual, where cost-effectiveness is a primary concern, closed circuit television cameras and monitor banks provide the most efficient security watch over bomb planting.

The basic internal security procedures sketched here will reduce a building's vulnerability to bomb attack and also provide substantial reductions in bomb searching times when investigating telephoned bomb threats.

VITAL AREA PROTECTION

The objectives of terrorism may be achieved through creating a state of psychological hazard, through inflicting actual casualties or deaths, or by structural damage to a facility. Should life loss, key personnel injuries or maximum economic or physical damage be the terrorists' objectives, then an attack will most probably be directed against vital parts of the organization.

Some examples of an organization's vulnerable vital areas are directors' offices, fuel stores, explosive/incendiary stores, communications centers, vital plants or machinery, computers and (in certain circumstances) personnel congregation points such as recreational facilities, lavatories and locker rooms.

External Threats

In certain cases, more especially in areas where terrorism has reached a mobile warfare phase, bomb security must be extended well outside the organization's perimeter. An ideal maximum life loss device might be planted in a vehicle near an organization's entrance, for example, designed to injure personnel reporting for work. In these and similar situations, parked vehicles should be under control near the facility's premises, and closely observed if they could present an explosive hazard.

The possibility of devices being planted to injure key personnel must also be considered, and preventive security advised and applied to their homes and personal vehicles as well as their offices. This concern applies to hand-planted and postal devices.

Upgraded Security

Areas which are particularly vital and vulnerable to bomb attack should receive upgraded security protection. This may be achieved by stricter key holder control, thorough body and luggage searching, regular security search sweeps, etc.

When developing vital area protection plans, security chiefs should first identify and classify these areas, analyze the risks and provide comparative security. Areas where explosion or fire would cause extensive damage or casualties should receive particular attention and, if necessary, be linked to area disaster plans with regional authorities.

STAFF DEPLOYMENT AND EDUCATION IN BOMB SECURITY

If an organization does not employ a security service trained to current explosive risk status, then security responsibility must be assigned to key staff.

As far as bomb threats are concerned, an organization will require a security supervisor, incident supervisors, plus sufficient personnel capable of supervising searching and evacuation operations. Management personnel with line experience may be used for the former supervisory tasks, and fire supervisors utilized to control searching and evacuation operations.

To provide continual security cover during shift changes, rest days and holidays, break periods, etc., a security rotation system should be developed to provide incident controllers at all times when the organization is operational.

Staff Training

All security and fire personnel should be trained in the response to telephoned bomb threats. The most effective training is achieved through a lecture room training program, closely followed by participation in simulated exercises. The loss in organizational productive time will be adequately repaid by increased efficiency and decreased time loss when handling actual bomb threat situations.

Switchboard and telephone operators, since they are the usual recipients of the telephoned bomb threat, must receive special instruction and training. Their ability to respond to the telephoned threat in a predetermined manner, without panic, will go a long way toward assuring a coordinated and efficient response to the bomb threat.

A Cooperative Effort

To achieve effective security requires the cooperation of all employees. Staff security awareness, cooperation and security consciousness are vital in achieving anti-bomb security. This cooperation is best achieved by an organization's management when their personnel feel that their part in the security program is both necessary and effective.

To eliminate the risk of intruders wandering through the facility, staff should be encouraged to check on and courteously escort strangers or visitors. To ensure that staff are motivated toward these precautionary measures, whenever visitors are brought into working areas they should be introduced to the senior staff member present. This helps to make the latter aware of the day-to-day routine and to provide a sense of security responsibility and corporate involvement.

Whatever security measures are adopted to minimize bomb risks, these should be explained to staff when this does not invalidate their effectiveness. This publicity, in the first place, reassures the employees and lets them know that security measures, even if inconvenient, are for *their* protection. Secondly, the publicity has a deterrent effect upon would-be intruders. As we have indicated, even the possibility of a one-in-five baggage check has a great deterrent effect on the typical bomb intruder.

In short, improved staff cooperation and active assistance will be better achieved when the necessary security is explained to them and made acceptable. This is particularly important when closed circuit television systems are utilized.

Chapter 3
HAZARDOUS DEVICES

It is not the purpose of this manual to go in detail into the construction of explosive devices. Unfortunately, as we have said, all too much information of this kind is available to potential bombers. It is important for security personnel, however, to understand the basic makeup of such devices in order to discover and recognize them more easily.

There are thousands of different hazardous explosive and incendiary devices which may be used by a bomber. They vary from a simple and often unreliable improvised pipe bomb to the most sophisticated booby-trapped device. The type of bomb usually found is relatively simple, but it can appear in so many disguises and with so many different ingredients that recognition is not always easy.

Because of the many variations in a device's construction, and because it may be booby trapped, *no device should ever be touched.* Bomb and incendiary disposal is a dangerous task, even for experts, and it must never be attempted by other than trained ordnance disposal experts. No matter how simple a device appears, it must not be touched or even approached by unqualified hands.

The Firing Train

Virtually all explosive bombs require a "firing train" to initiate an explosion. The firing train consists of a detonator, booster and main explosive charge.

The firing train starts with the detonator. As this burns or produces a small explosion, a booster charge is exploded. This in turn produces sufficient heat and energy to explode the main charge. There are, of course, many variations on this firing train, due to the differing properties of explosives and detonators.

Explosive Charges

A bomber may use many different types of explosive. Explosives may be improvised from household or agricultural chemicals, or stolen from mining stores, military stores, etc. Additionally, common gunpowders (nitro powders and black powders) may be obtained from retail sources or stolen.

Because of the variety of commercial and improvised explosives that may be encountered, any unknown solids, powders, crystals, liquids or sludges should be treated as suspect, especially if they appear to be connected to a firing train.

Low Explosives

Black powder, nitrocellulose and smokeless powder are common examples of low explosives. These are principally used in small quantity as propellant powders, as in loading ammunition.

Black powder, because of its ready availability and lively combustion, is a good example of a low explosive which is frequently used in improvised hazardous devices, such as pipe bombs. Black powder is very sensitive and can even explode from the influence of static electricity. In terms of blast damage, black powder, like other low explosives, produces a low shattering effect, although it exerts a strong pushing force. Exploding in any large quantity, black powder is capable of pushing loose furniture and other articles for some distance.

Black powder consists of small, flat, shiny black or gray flakes of powder. In a crude but dangerous form it can be improvised from a mixture of common chemicals. In improvised form it will appear as a powder or crystals of black, yellow and white.

A black powder device may be built around any container, such as a gunpowder tin, pipe, etc. Because of its high density, it can also "explode" with some force without being contained.

PULL DEVICE

TRIP WIRE

OPEN CONTACTS

ELECTRICALLY INITIATED EXPLOSIVE

A basic trip wire activated device. When trip wire is pulled, the electric contacts on the clothes peg snap shut and complete the circuit.

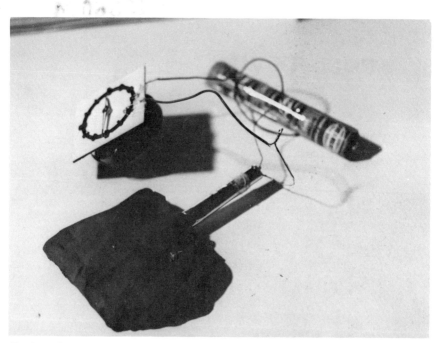

Clockwork time delay, with electrical detonator, battery, clock and plastic explosive.

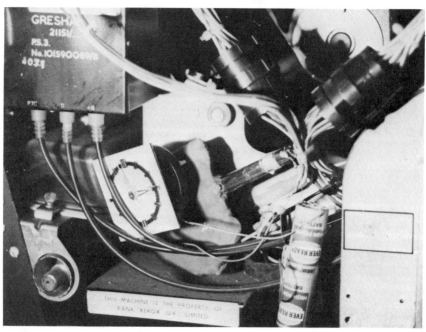

Clockwork time delay plastic explosive device used for machine sabotage.

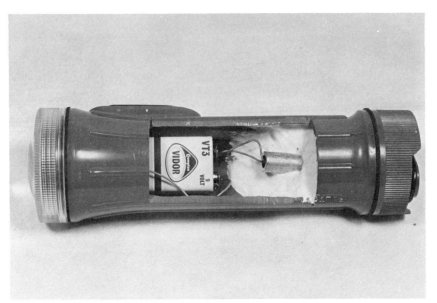

A flashlight device, which detonates when switched on.

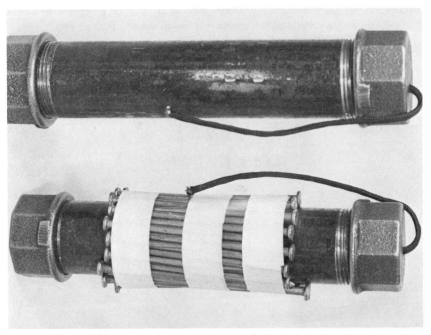

Pipe bombs, packed with explosive (often black powder) and fitted with a pyrotechnic fuse. These are common and easily improvised devices, often surrounded with nails (lower portion of photo) or mud-packed stones to increase their anti-personnel effect.

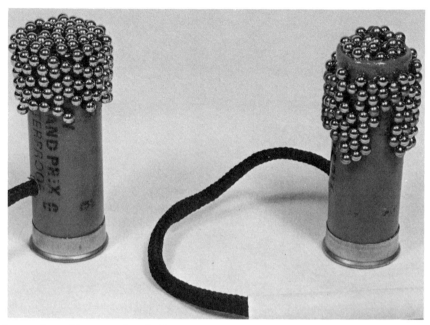

Improvised 12 gauge shotshell grenade. Similar percussion devices with booby trap releases may use cartridges inserted into gasoline cans, bottles, etc.

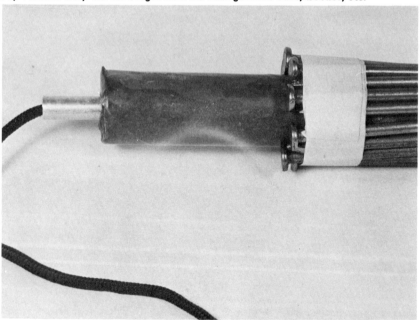

A simple improvised device consisting of a pyrotechnic fuse and detonator, activating a stick of dynamite, surrounded by nails to increase its effect.

High Explosives

A wide range of commercial high explosives are available, in many forms and variations. Additionally, many dangerous and unstable high explosives can be improvised from easily available chemicals.

The common high explosives consist of such substances as T.N.T., dynamite, nitroglycerin, and plastic explosives.

T.N.T., or trinitrotoluene, is usually formed in creamy yellow blocks, weighing between one-half and one pound.

Dynamite is formed in sticks, resembling short candle sticks wrapped in greasy brown paper. As dynamite ages, it forms droplets and crystals of highly dangerous nitroglycerin on its surface. This is particularly likely to happen when dynamite is stored in hot or damp areas. When nitroglycerin forms on dynamite, the sticks can explode from the slightest vibration.

Nitroglycerin is an odorless, thick, milky liquid. It is a very sensitive explosive, and becomes even more dangerous as it ages. As nitroglycerin becomes older, it turns to yellow-green, and is very unstable.

Plastic explosives are produced in slabs, flat sheets, tubes and rolls, etc. They can be moulded around an innocent-looking object, with a blasting cap pushed into the surface. Many variations of plastic explosives are commercially produced, and only small quantities are necessary for anti-personnel effect. Therefore, any substance fitting the above approximate description should be treated as suspect.

High explosives produce far more shattering effect than low explosives. Therefore more structural damage is achieved, and secondary hazards from flying missile debris and broken glass become more dangerous.

Incendiaries

As with explosive devices, incendiaries are widely used to cause damage, panic and injuries. Annual fire losses through structural damage and content losses are enormous, and the risks of incendiary devices planted for criminal or terrorist gain must be countered effectively. Incendiary devices are often more common than explosives, in some States outnumbering explosive devices by 3 to 1.

An incendiary device can be as simple as a match tossed into a trash can or as complex as a delayed burning chemical device. Quite a number of easily available agricultural and household chemicals may be used to make improvised incendiaries.

Any container with chemical mixtures, such as powders, solids, sludges or liquids, especially if connected to flammable materials or an improvised fuse, should be treated as suspect.

Gasoline and other fuels, when contained, are not only flammable but have an explosive effect almost equal to that of high explosives. Gasoline type devices, on automobiles or elsewhere, are therefore very dangerous.

ACTIVATING EXPLOSIVE DEVICES

Although only explosive ordnance disposal experts should attempt the rendering safe of hazardous devices, some knowledge of device activation methods will again assist security staff in the recognition of potential devices.

There are at least a score of ways of activating explosive devices, and a combination of several methods may be used in one device. Some of these methods include heat, light, sound, magnetism, vibration, electrical contact, chemicals, release of tension, increase of tension, radio waves, mechanical systems, etc.

For the purposes of our tactical situation we can divide hazardous devices into two types: time delay devices and antipersonnel handling devices.

Time Delay Activations

Time delay devices, activated by several different systems, are most frequently used in "precise" bomb threats. Here the bomber sets his device to activate at a precise time, and warns (or does not warn) the organization with a bomb threat.

Time delays on hazardous devices may be achieved by mechanical or chemical means. In the former case, a watch or clock mechanism is normally used, although any motor system will be suitable. Chemical systems usually incorporate two chemicals which, when combined together, ignite after a calculable time lapse.

Mechanical devices, built up from watch or motor mechanisms, are often connected to electrical systems. An electrical circuit is held open by contacts separated around the clock face, but as the hands make contact the circuit is completed and detonation occurs. Such a circuit can be improvised from a wrist watch or an alarm clock, or even a wall clock in the facility under attack.

Chemically activated devices are usually built in test tubes or cigar tubes. Apart from chemical mixtures, acid mixtures that dissolve electrical contact wires, and other systems, the activation can be achieved by something as simple as a tube full of wet rice expanding and pushing an electrical contact upwards to complete a circuit.

Incendiary devices usually have time delays, which could be as described above for mechanical or chemical systems. Usually the delay is a long burning fuse or chemical mixture. The fuse may be of commercial manufacture or improvised from impregnated rope or candles.

Anti-Personnel Handling Activations

When we enter the field of booby traps, the devices and activation methods are limited only by the skill and ingenuity of the bomber. Thousands of different devices may be improvised, although basically they can be compartmentalized into quite definite sections because of the limitations of explosives and initiators.

The essence of booby trapping is "infinite variety and low cunning." Within the tactical parameters of this manual, therefore, it is possible to provide no more than a brief warning as to the presence and possible types of booby traps. More detail may be found in Appendix A.

Although many other activation systems or combinations may be used, a booby trap is usually constructed of a mechanical or electrical activation system, or a combination of both these.

Mechanical systems include the use of trap wires, release of tension wires, push/pull switches, vibration switches, mouse traps and clothes pegs, door hinges, etc. Electrical activations include switches of all types, skin contact sensitive circuits, push/pull switches, pressure switches, increase and decrease of charge circuits, and many others.

Any out-of-place or dumped object should be treated as suspect in the search for a booby trapped bomb. Arms finds, narcotics or contraband, or anything likely to be quickly handled or picked up out of curiosity, are ideal traps. Clues to the presence of a bomb might be pegs, nails, pieces of cut wire, thin string or nylon lines. Evidence of disturbed ground, sawdust or loose floorboards may indicate the presence of a buried device. Replastered walls, brick dust or loose electrical fittings might indicate that a wall switch device has been planted. In general, anything that doesn't belong where it is found, or which appears to be altered or disturbed, should be treated as suspect.

Blasting Caps

Many electrical devices use a blasting cap to detonate their main charge. There are two types of blasting caps, electrical and pyrotechnic.

Electrical blasting caps are easy to recognize, usually being bright metallic, cylindrical capsules of copper or aluminum, closed at one end. They are about the diameter of an ordinary lead pencil and can be up to six inches long.

Great care must be taken not to handle or disturb electrical blasting caps, because only a small amount of electrical charge is required for detonation. Even static electricity can detonate a blasting cap, which itself contains a potentially deadly explosive charge.

Radios should not be used near electrical blasting caps or electrically sensitive (radio sensitive) explosives, since radio energy can cause a charge to explode.

Detonation from low wattage communications radios is unlikely, but for safety a 50-meter safety radius should be maintained from suspected devices. Radio *transmission* is infinitely more dangerous than reception modes, which are unlikely to activate a device—unless it has been set to detonate at that radio's particular frequency.

Pyrotechnic blasting caps (non-electric) are sometimes used in devices, either of commercial manufacture or improvised. Non-electric blasting caps usually have a colored pyrotechnic fuse leading into one end. The fuse can be a variety of colors, depending on manufacture and burning speed.

A less obvious device, with a mouse trap activator. When can lid is lifted, mouse trap snaps shut, contacts on either side touch and complete the circuit.

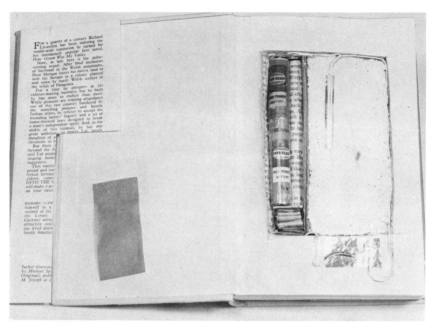

Book, which appears normal when closed, contains explosive in cut-out compartment; detonates when cover is opened.

Gasoline can incendiary device, with improvised pyrotechnic fuse made from drinking straws.

SAFETY PRECAUTIONS

It must be stressed once again that handling suspected bombs and attempting to render them safe is a task for specialists. No matter how simple it appears to be to cut a firing train, no suspected device should ever be touched or approached too closely.

The device itself may be booby trapped, and its approach route may also be booby trapped with floor pressure switches or trip wires. During training, on simulation exercises and on actual alerts, all involved personnel must be cautioned not to touch or try to approach a suspect device.

Devices that fail to explode are also hazardous. They may be set as bluffs, booby trapped so as to detonate when handled by the unwary. Or the reason for their failure may be a minute particle of dirt on an electrical circuit, which will be freed if handling is attempted. Time devices using clock or watch mechanisms may fail to make proper contact, and yet make contact after another twelve hours on the next sweep around the clock face.

Consequently, no matter what the type of device, its apparent simplicity or apparent failure, it must not under any circumstances be approached or handled by unqualified personnel. Remember the catch phrase:

Touching Triggers Tragedy!

Chapter 4

TELEPHONED BOMB THREATS

Excerpt from FBI Uniform Crime Reports:

BOMB HIGHLIGHTS — 1975

Thirty-nine percent of all bombing matters occurred in cities over 250,000 population. California led the Nation in number of incidents with 533 or 26 percent of the total. The Western Region experienced 789 or 38 percent of the 2,074 bombing incidents.

Seventy-six percent of all bombings occurred between 6:01 p.m. and 6:00 a.m. Eight hundred and eighteen occurred between 6:01 p.m. and midnight, and 752 occurred between 12:01 a.m. and 6:00 a.m.

Monday had the highest frequency of bombing incidents with 338, while Thursday had the lowest with 279.*

As the F.B.I. report on bomb damage during 1975 indicates, bombings and bomb threats can occur at any time, any day, and in any place, in cities large or small.

How are such threats initiated? In most cases, they are made by telephone, usually to an organization's switchboard but occasionally on other extensions. While only about 2 percent of reported bomb threats turn out to be real, there is no way of predicting whether a threat call will be a thoughtless prank, an irresponsible ruse to cover staff returning from an extended alcoholic lunch break, a deliberate method of harassment designed to cause disruption or panic—or a warning of an impending explosion. But because the potential risks

*Bomb Summary, p. 4.

of injury to persons are so high, few threats can be ignored.

Actual bomb threats may be made anonymously by the bomber himself. However, an anonymous call may also be received from someone other than the bomber or bomb planter. An innocent person may see a suspect device and make such a call, refusing to give his name so as not to become caught up in a lengthy police investigation or publicity.

If the call is genuine, then its motivations for the actual bomber are to minimize casualties or, occasionally, structural damage. The bomber's aims of harassment or political attention are usually achieved without causing injuries—which can sometimes be counterproductive. There are, of course, many other possible motivations for warning calls, but the most common one is a desire to give the organization time to evacuate people and avoid injuries.

Because personnel injury is not often the primary aim of the bomber, explosives will often be planted to detonate when an installation or facility has fewest staff present, usually at night. In such cases the bomber will tend to place his device in mid-afternoon, and security should be tightest during these periods. Because these devices are set to detonate during quiet or off-duty periods, threat calls received at such times are often serious.

Also—again because of the desire to minimize casualties—serious threat calls are usually very specific, and they may even be repeated in a few minutes.

TELEPHONE BOMB THREAT PROCEDURES

Bomb threats are usually received by a telephone operator on a publicly listed number. These few seconds will probably be the organization's only contact with the caller, so it is important that any available information be recorded accurately. If the telephone operator is trained in the handling of bomb threat calls, she will be able to provide the security service, or police, with vital information. If the operator is untrained, the call message will all too often be repeated grossly inaccurately, and chaos may result.

There are two essential points of information which the telephone operator must discover. These are:

1. The expected explosion time.
2. The bomb's location.

For investigative purposes it is important to find out as much as possible about the caller. Conversation should be gently encouraged, because the longer the caller is on the phone the greater are the chances of tracing him via the telephone company. Also, by engaging the caller in conversation the operator may learn more about him or hear clues that will be useful for later investigations. This applies not only to the caller's regional accent or race, but also to background noises, such as the sounds of road repairs, traffic or trains, machinery, music, conversation, etc.

However, telephone operators should be trained not to upset or panic a bomb threat caller. Asking too much for too long can make the caller nervous and cause him to hang up. The essential information is the time and place of the explosion—get this information and then, and only then, try to get more.

Using a tape recorder linked to the switchboard, either permanently recording or activated by an emergency switch, is the best way to aid investigation and analysis of bomb threat calls. Telephone operators cannot be expected to report accurately on a bomb threat call of a few seconds duration. Where it is legally permissible, the use of a tape recorder is strongly recommended, certainly when there has been a history of bomb threats.

When bomb threat calls are a common hazard, arrangements can be made through the telephone company's security department to have calls traced rapidly. This, plus the effective prosecution of threat callers, can act as a deterrent to frequent hoax calls.

Operator Training in Bomb Threat Procedures

An effective security response system can be developed to handle telephoned bomb threats. But it must be remembered that the threat starts on the telephone. The response of the operator receiving the call must be the strongest link in the security chain.

Training for telephone operators in handling bomb threat calls should include both a short period of classroom instruction (preferably utilizing visual aids) and occasional simulated exercises. It is not enough simply to describe what should be done. As with fire drills, immediate and effective response to a real situation can best be attained by the "experience" of having gone through the procedures, even if only in mock drills.

Minimum training should cover at least the following areas:

1. *Remain calm.* If a telephoned bomb threat is received, try to keep the scene quiet and normal. Even though 98% of threat calls turn out to be hoaxes, the company has a special security plan to handle the situation. The staff's safety is considered uppermost in that plan.
2. *Follow pre-planned procedure.* If a threat call comes in, immediately follow a step-by-step procedure such as the following:

- Switch on tape recorder (if available)
- Press emergency switch for security assistance
- Record time of call
- Record bomb's location
- Record time of explosion
- Inform security service/police

So that all of this can be remembered in the stress of a real call, a simple form should be provided and should be kept at the switchboard or operator's desk, such as the one shown in Figure 5.

BOMB THREAT PROCEDURE:

Time of Call: _____

Bomb Location: _____

Explosion Time: _____

Call SECURITY _____ EXTENSION _____

If no reply within 1½ minutes, call POLICE EMERGENCY

Stay where you are.

Figure 5. Telephone operator's bomb threat procedure form

3. *Get the two vital facts*—the bomb's location and the explosion time. As soon as the caller starts talking, start

TELEPHONED BOMB THREAT REPORT

Supposed Bomb's Location

Supposed Explosion Time

Time of Phone Call

Phone Booth (Call Box)　　　　YES　　　　NO

CALLER'S EXACT WORDS:

Caller's Sex　　　　　　　　MALE　　　FEMALE

Caller's Age　　　　　　　　CHILD / YOUTH / ADULT

Caller's Mental State　EXCITED / NERVOUS / IRRATIONAL / CALM

Speech / Voice

Regional or Other Accent

Any Background Noises?

Number Call Was Received On

Operator's Actions After Receiving the Call:

Any Additional Information:

Operator's Name _____

Home Address _____

Home Telephone Number _____

Date of Report _____　　　　Time _____

　　　　　　Signature _____

Figure 6. Telephone operator's bomb threat report form

SERVICE	OFFICE PHONE EXT.	RADIO CHANNEL	EMERGENCY NO.	HOME NO.
Security Chief				
Security Office				
Emergency Security				
Fire Chief				
Fire Office				
Fire Emergency				
Health Officer				
Health Center				
Emergency First Aid				
Elect / Eng Maintenance				
POLICE EMERGENCY FIRE EMERGENCY AMBULANCE " HOSPITAL " DISASTER " TELEPHONE CO " ELECTRIC CO " GAS CO. "				

Figure 7. Emergency contact list

filling out the procedure form. The more information that can be discovered the better, but the two key facts are paramount. Let the caller talk at first, but if he finishes and has not given the explosion time and location, gently try to find out more. You can pretend that you cannot hear, or that there is a poor connection with crackling on the line. If the caller does not want to talk, and has not provided the essential information, say "Don't hang up" and repeat the questions. Showing some disbelief might goad the caller into giving more information, but be careful not to provoke him too much or he may hang up prematurely.
4. *Encourage him to talk.* The longer the caller can be kept on the line, the better. Keep listening to everything he says, and pay attention to any background noises. If the caller remains on the line, ask him why he has planted the bomb. He may want to talk about some emotional problem, grievance, or political idea. Your sympathy and interest may hold the caller on the line long enough for the call to be traced, or encourage him to give essential information.

Once the telephone operator has carried out the emergency procedure described and the call has been terminated, the security department and police will carry on and deal with the situation. The operator's task is not yet finished, however, for that initial call will often be the organization's only direct contact with the bomber or hoaxer. As soon as possible, while details are fresh, a more detailed bomb threat report should be completed, such as the one shown in Figure 6.

As an aid not only in handling bomb threat calls but also for other emergencies, emergency contact information such as that in Figure 7 should be kept by every switchboard. A similar, but possibly more detailed, copy will be required at the security control center.

Chapter 5

BOMB THREAT SECURITY RESPONSE

Every organization should have a basic security response procedure to deal with typical bomb threats. The response systems outlined in this chapter may be used by conventional security services or, in smaller organizations, by management personnel. In all situations, and on every shift, there should be someone responsible for the planned security response to such a threat.

Upon notification of a bomb threat from the telephone operator receiving the call, the Security Director, Control Room Supervisor (if such a central security station exists), or the security officer in charge on a given shift should immediately dispatch the senior security guard available to the supposed bomb site, where he will act as Incident Commander.

The Security Director should then question the operator who received the call. The following investigation should be made to acquire the necessary information for organizing an appropriate security response:

1. When was the call made?
2. Where is the supposed bomb's location?
3. When does the supposed bomb explode?
4. What were the words used by the caller? Did he sound serious? Was there muffled or childish laughter in the background? (The latter conditions would probably indicate a hoax call.)
5. Have there been many previous hoaxes?

6. What danger does the threat create for personnel or resources?

This information should be quickly investigated and evaluated. Ideally, policy decisions can be made and relayed to the Incident Commander while he is on his way to the reported bomb site. In any case of doubt, the threat must be treated as actual.

At this stage there are five alternative security responses that can be made:

A. Immediate *full* evacuation.
B. Immediate *localized* evacuation.
C. Personnel work area search.
D. Security personnel search.
E. Police search/police control.

Each of these alternatives is discussed in the following pages.

The decision to evacuate is ultimately the responsibility of management; it cannot be delegated to a search team, be it police, fire or internal. Search personnel or a bomb technician may make recommendations to management regarding evacuation, but it should be emphasized that the only thing certain about explosives and explosions is their uncertainty. Too many factors are involved for anyone to predict with absolute certainty how much damage a given explosive charge will create in a given area. In such circumstances, only management can make the final decision to accept the risk a bomb threat creates, however improbable or remote that threat may seem to be.

In organizations where a trained, professional security staff is employed, and particularly where the Security Director is given management level responsibility, swift response to a bomb emergency suggests that he should have the authority to evacuate in the face of a threat. At the very least he must have immediate accessibility at all times to a management representative with the authority to make that decision. There may be little time for discussion, and decisions must often be made quickly. *It is vital that the authority to decide when to evacuate be clearly determined and understood in advance of any bomb threat decision.* Pre-planned emergency responses should not be countermanded during a crisis by management or other personnel, who should act only as local advisors.

Bomb Threat Security Response   57

In any event, key managers should be informed of the situation as early as possible. This may help in maintaining a calm atmosphere and also prevent the access of unnecessary people into hazardous areas.

If the threat creates a serious risk, the Security Director should immediately place on alert any local disaster plans, as well as fire, ambulance and police services.

At this early stage of the response operation, it is important that the security department does not over-respond. An excessive response can cause panic and confusion, a problem compounded if—as will probably be the case—the threat turns out to be a hoax.

In installations where vital resources are being held, it should be remembered that a bomb threat may be only a decoy or diversion for a robbery attempt. If this is a possibility, then physical security should be reinforced at key locations.

Where vital resources protection is of paramount importance, police and other local authorities called to the bomb threat situation should be checked for positive identification before admittance to the area is granted. In such cases uniforms or official looking vehicles cannot be accepted as I.D. Despite some time losses, positive I.D. checks should be made.

An innocent-appearing bicycle may have its tubular frame packed with explosives.

A car park beside an office block. If possible, such areas should not be used during evacuation because of risk of flying glass. All assembly areas should be searched before evacuation. Likely places for devices to be planted are flower bed, parked vehicles, exterior lights, trash can.

Security Responses

The security response will depend upon the evaluation of the threat call and the time available. It should be noted, however, that the timing mechanisms on improvised devices are often crude and inaccurate; therefore an allowance of fifteen minutes should be made on the supposed explosion time wherever possible.

Responses may be any, or any combination, of the following:

A. *Full Evacuation.* Full evacuation is not, in fact, often essential. It should only be used where explosive risks are very high, or positive proof of a large device exists.

B. *Localized Evacuation.* Where there are reasonable grounds to believe that a small explosive device may exist, or casualty risks can be minimized by evacuating a small area, localized evacuation should be practiced. This can be done without creating disturbance in surrounding areas in non-hazardous zones.

C. *Personnel Work Area Search.* If the situation is seemingly routine and does not create excessive risks, work area supervisors should have their personnel make a localized search. They should be able to recognize quickly any strange or out-of-place objects in their work areas.

D. *Security Personnel Search.* Concurrent with any of the above three responses, security officers may search potential bomb planting sites and take control of searching in lower risk areas.

E. *Police Search or Control.* In certain circumstances, the organization may wish to pass complete or partial control of the bomb threat situation over to local police units. In this event, the police may take over searching, utilizing screening equipment and other emergency controls. Security personnel and other personnel supervisors should then be seconded to assist police units.

EVACUATION

Full or localized evacuation is a common response to a bomb threat. Basically, evacuation should be used only as a last resort— when there is sufficient evidence that a device may exist. Evacuation is rarely an immediate necessity in bomb threat situations. However, evacuation injuries, caused through panic or confusion, can be as dangerous as an explosion. Therefore an evacuation plan should be developed and practiced, so that in an emergency an orderly evacuation can take place without confusion.

The greater the number of people that must be evacuated, the greater is the possibility of panic casualties. Panic spreads quickly, like a contagious disease. An over-hasty evacuation of a large crowd can cause a near riot within a few minutes. Catastrophic injuries have resulted in these situations—often far worse than those following a small explosion.

Not only does an over-hasty evacuation create the risk of likely casualties, it also extends the time taken on a bomb search. With a safe, methodical evacuation plan in operation, a building can be cleared very quickly.

It should be noted that bomb threat evacuation differs considerably from standard fire alarm evacuation. Different security, fire and general personnel training and warning notices will accordingly be required.

Clearing Evacuation Routes and Assembly Areas

During an evacuation, large numbers of people must pass through public or commonly used corridors, stairways, etc., and then assemble in a large group. It is essential that during this process personnel are not exposed to even greater explosive risks. These public areas are potential bomb planting sites, and a device exploding in the midst of an evacuating crowd can cause devastating casualties. To avoid this possibility, these potential danger spots should be quickly searched prior to any evacuation.

Another potential risk is that evacuation assembly points will be clearly displayed on fire and other safety notices. A devious bomber could achieve maximum anti-personnel effect by placing a device at an assembly point. Although this occurs infrequently, except in specific terrorist actions, it is a high risk potential. Evacuation assembly points must therefore be searched prior to evacuation.

Evacuation Warnings

The evacuation method will depend upon the tactical situation. Where only a small number of people have to be controlled, then instructions can be given in person. This is preferable in avoiding any panic. Where a large number of people must be controlled, evacuation instructions should preferably be given by supervisors or security officers.

The least desirable system is to make an announcement over a public address speaker. However, when this remote method is used, full advantage should be made of the speakers to provide complete and detailed instructions for a safe evacuation.

Discreet Warnings

When a public address system is used, or even when instructions are given personally, it is not always necessary to announce the nature of the bomb threat. People may be evacuated by telling them that the building has a gas or water leak which must be repaired, or that the air conditioning ducts are leaking unpleasant fumes and must be cleaned. This is an excellent means of minimizing panic and worry.

When information must be passed over a public address system, with a wide area coverage, but it is undesirable to alert general personnel, then a discreet or coded message may be used. For example, the announcement might read, "Would Mr. Smith from ASD Inc., please meet Mr. Sandle in the fuel store." Such a coded message could alert security guards of a bomb threat in a specific area. Using such systems provides security against unnecessary people knowing of the emergency, and also prevents a bomber still within the premises from knowing that a security response is in action.

Use of Fire Alarms

Fire alarms may be used for a bomb threat evacuation warning, if speed is essential and no other system exists. However, standard fire alarm procedures can often increase the hazards of an explosion. Such procedures include, for instance, the shutting of all doors and windows. If an explosion occurs, this will increase the blast and potential casualty effects. Fire alarms should therefore not be used unless a very fast evacuation cannot be achieved by other methods.

If a bomb search is made after a building is evacuated by means of a fire alarm, then (where tactical conditions permit) doors and windows should be opened prior to searching. Obviously this involves a loss of time that could better be used for the actual bomb search.

Evacuation Scales

If a device is situated in a non-vital area, i.e., not near flammable or explosive stores, then the *minimum* evacuation distance should be 100 meters for a small device and 200 meters for a medium-sized device. It must be stressed that these are minimum figures; they may have to be considerably increased, depending upon the tactical situation. Figure 8 provides more detailed recommended evacuation distances for both open and enclosed areas. Again, no chart can take into account all of the potential factors involved.

If the position, type and size of a device are known, then a full building evacuation may not always be necessary. The decision as to the type and scale of evacuation necessary is mainly in the hands of the Incident Commander, referring to his Security Chief as necessary for policy.

Sometimes it is necessary only to evacuate a couple of rooms on either side of a suspected device, and a story above and below. If a security search or police Bomb Squad advice suggests that a greater threat exists, then evacuation scales can be increased. Basically, the smaller the evacuation the better.

Crowd control can be a problem in bomb threat situations, because people's curiosity often far outweighs their common sense. Thus, strict cordon lines should be maintained at a safe distance from a danger zone, and security guards or supervisors should be used to keep people clear.

Whenever possible, personnel should be evacuated behind solid cover to protect them from possible blast and missile effects.

Bomb Threat Evacuation Methods and Instructions

1. All personnel should leave through main or fire exits, in a quiet and orderly manner. A walking pace should be used. Running will only cause panic and injuries.
2. Personnel should assemble in the prearranged safe areas after evacuation. Supervisors will then hold a roll call to ensure that all personnel and visitors are accounted for. Any not present should be immediately reported to Security.

RECOMMENDED EVACUATION DISTANCES

A. For persons in exposed areas:

Explosives (Pounds Weight)	Safe Missile Distance (Feet)
1-27	900
30	930
40	1020
50	1104
60	1170
70	1230
80	1290
90	1344
100	1393
150	1593
200	1754
250	1890
300	2008
350	2114
400	2210
450	2299
500	2381
550	2458
600	2530
650	2599
700	2664
750	2725
1,000	3009

B. For persons in enclosed areas (inhabited buildings):

Explosives (Pounds Weight)	Safe Missile Distance (Feet) Unbarricaded	Barricaded
2 - 5	140	70
5 - 10	180	90
10- 20	220	110
20- 30	250	125
30- 40	280	140
40- 50	300	150
50- 75	340	170
75-100	380	190

Figure 8. Recommended evacuation distances for bomb threat incidents

3. As they leave, supervisors should try to disconnect any electrical apparatus, such as calculators, fans, heaters or typewriters.
4. Building custodians should switch off major plant equipment, such as air conditioning, heating, extra lighting, etc. To prevent time loss in an evacuation, this may be accomplished later by remote switching from a master console.
5. Elevators or lifts should *not* be used during evacuation.
6. As personnel evacuate, they should be instructed to open all possible doors, windows, cupboards, filing cabinets, etc.
7. As personnel evacuate, they should be instructed to take their personal belongings with them, providing that this can be accomplished quickly.
8. Supervisors, who should leave the area or the building last, should try to open as many doors and windows as possible.

It is recommended that a basic list of evacuation instructions, such as the one shown in Figure 9, be issued or displayed to all personnel for bomb threat situations.

BOMB THREAT EVACUATION PROCEDURES

1. Walk out of the building in a quiet manner. Do not cause other people to panic by running.
2. Go to _____ and wait for instructions.
3. Do *not* use elevators when you evacuate.
4. Leave drawers, doors and windows open.
5. Take your personal belongings with you, if they are nearby.

Figure 9. Bomb threat evacuation guide

SEARCH OPERATIONS

All bomb threat situations will require a search operation. There are three types of bomb searches, two requiring internal personnel and the third involving external law enforcement units. These three search types are:

1. Employee search (immediate work areas).
2. Security personnel search (mainly potential bomb planting areas).
3. Police search (utilizing mobile detector apparatus, dogs, etc.)

An employee search of their immediate work areas is probably the first and most efficient response to a bomb threat. Only employees will be able to determine at a glance whether many objects are harmless or potentially dangerous infiltrated devices. They can also quickly detect any strange or displaced items in their normal working environment. Security staff cannot perform this function as quickly and reliably.

Where it is compatible with safety—i.e., serious doubt exists about a device's actual presence—or where there is sufficient time for searching, women should also search their work areas. Ladies' rest rooms and similar areas should also be searched by women staff.

Beyond this employee search of their work areas, on an initial response to a bomb threat situation the organization's security officers should search such areas as reception lounges, corridors, rest rooms, locker rooms, lavatories, etc. Their specialized skills in the recognition of devices are required in these high-risk areas. After these areas have been searched, security personnel can move on to search general areas and other potential targets. For example, since a device is likely to be planted in a place which is fairly easily accessible to the bomber, such as the exterior of an area or building, these areas should receive full attention in the bomb search.

Lastly, should these two searches prove negative but the situation suggests that the organization's risk status requires further investigation, local law enforcement services should be utilized. The police Bomb Squad or Hazardous Devices Squad will then be dispatched to take control of the operation. Such police units are equipped with remote control detection and examination equipment, explosive detecting dogs, etc., and are best utilized to make clearance searches and to positively identify suspicious objects.

Search Safety

Even though, as we have said, only about 2% of the thousands of bomb threats received may be actuals, searching must still be considered a hazardous task. It must therefore be conducted with the maximum attention to safety. The search methods outlined here have been developed to minimize the possibility of casualties and to provide maximum speed in the search. A fast search is a safe one—and it reduces production loss time.

Searching should not be thought of only in a crisis. It should be practiced by both general employees and security personnel, preferably during simulated exercises. Only in this way can the necessary safety procedures be taught, and speed in searching acquired.

The following basic search safety rules must be strictly observed in *all* searching operations:

1. Never use more searchers than absolutely necessary.
2. Use a maximum of two searchers per room, or for an area up to 250 square feet.
3. Use searchers in alternate rooms or areas.
4. Never assume that only one device has been planted; continue searching operations until the whole area is cleared.
5. Clearly mark and report areas searched and cleared.
6. Clearly mark and report areas found hazardous.
7. Do not allow searching for more than ten minutes without a few minutes rest.

Each of these rules is designed to minimize casualty potentials. Keeping the number of searchers to the minimum reduces the number of persons within a hazardous area should an explosion occur. The same principle governs the use of searchers in alternate rooms and areas, as well as the deployment of only two searchers within a large room or area. These two should work widely separated, each searching half of the area. More searchers will not only create additional hazards in the event of an explosion, but will also result in less thorough searching.

Planting two or more devices, set to explode when personnel are re-occupying the premises, or while police are investigating, is a common ploy of bombers. Thus, searchers should *never* assume that only one device exists. Searching must be continued until the threatened area is completely cleared.

Areas which have been searched and cleared should be so

marked on the door or another conspicuous place with adhesive labels, tape or chalk. If suspect devices are encountered, or suspicion is raised about an area, it should be clearly marked with a sign such as "DANGER — BOMBS."

Lastly, searching is an exacting and tiring operation. To be effective, it requires 100% concentration. Consequently, after they have been searching for about ten minutes, personnel involved should be given a few minutes rest. This will help them to maintain their caution and concentration when the search is resumed.

Search Techniques

The first two stages of a search operation—employee work area search and security personnel search—should be conducted under the supervision of senior security officers. This is to ensure that search safety rules are followed.

The purpose of searching is to detect and report hazardous devices to the police bomb squad specialists. It is *not* to touch devices or even to approach them too closely. Such tactics are renowned for placing "heroes" in wheelchairs, and they serve only to endanger the lives of everyone involved.

A room or area should be searched slowly and systematically. To avoid the possibility of any areas being missed, the room should be split into two halves, and each should be searched separately. Each half is then searched in three sections: floor to waist level, waist to eye level, and eye to ceiling level.

As the searcher enters an area, he should move slowly and carefully. Speed in the search comes from organized search procedures, not from hasty or reckless movement. Booby traps or trip wires may have been planted, and the searcher should be wary of these. When entering the area, he should pause and listen. In this way he may be able to detect ticking clockwork devices. (This is one of the reasons for switching off power equipment during evacuation.)

The basic principle in bomb searching is to *trust nothing and assume nothing is safe*. Searchers should remain fully alert, for a bomber can conceal a device in almost any innocent-looking article. Sanitary towel dispensers, lavatories and cisterns, excreta, dead bodies, motor or other accident victims, pornographic books and telephones are just a few of the possible bomb planting covers.

BOMB SEARCH REPORT

REMEMBER: Keep well spaced apart.
Do not search in adjacent areas.
DO NOT TOUCH SUSPICIOUS OR OUT-OF-PLACE OBJECTS.

Area Searched _____
Suspicious object found at _____
Description of object _____

Sketch of object

Did the object have a visible or ticking clock timer?   yes/no
Did you see or smell a burning fuse?   yes/no
Did you see trip wires or booby traps?   yes/no
At exactly what time did you find the object?   _____

Sketch of object's position

Is the object easily accessible?   yes/no

DRAW YOUR EXACT ROUTE TO AND FROM THE OBJECT

Did you continue searching the area for other devices?   yes/no
Did you see other suspicious objects?   yes/no
Has the area around the object (above/below) been evacuated?   yes/no
Are there people near the object?   yes/no
If yes, how many people are there?   _____
Are all evacuees accounted for?   yes/no
Is the suspicious object near: *fuel/explosives, valuable equipment?*
HAND THIS FORM TO THE BOMB SQUAD WHEN THEY ARRIVE.

Name: _____ Rank: _____

Figure 10. Bomb Search Report

Elsewhere in this text we have provided some details on the recognition of fuses, detonators, explosives and incendiaries. (See Chapter 3.) These clues to the recognition of concealed explosive devices bear repeating, along with some other things to look for, such as:

- Recently disturbed ground
- Sawdust
- Brickdust
- Wood chips
- Electrical wire
- String
- Fishing line
- Dirty rope (fuses)

- Tin foil
- Partly open drawers
- Fresh plaster or cement
- Loose floorboards
- Disturbed carpeting
- Loose electrical fittings
- Out-of-place objects
- Greasy paper wrapping, etc.

Once an area has been found safe, it should, as indicated above, be marked "O.K." or "SAFE." Similarly, should a suspect device be reported, the area should be marked "DANGER — BOMBS" and evacuation control should be stepped up.

When a suspect device is reported to the senior security officer at the incident site, he should obtain an accurate idea of its exact position and description. If possible, a plan of the device's appearance and position should be drawn. This information can then be reported to the police bomb squad, to whom it can be invaluable.

Minimizing Explosion Damage

In some cases it is possible to reduce the damage caused by a device exploding. In vital areas, where lives or critical material are at risk, it may be expedient to attempt to reduce possible explosive damage.

Anti-explosive force foams and blast suppression blankets may be used to cover a device. In an emergency, sandbags may direct blast away from critical areas. It is essential, however, that no hard shield be placed near the device, for these will increase the blast danger effects considerably.

As pointed out in our discussion of evacuation, damage and casualties will be considerably reduced if all possible doors and windows are opened. This allows the explosive force to be quickly vented. Utilizing plastic coating or blast-suppression curtains will also reduce damage and casualty potentials.

Vehicle Searching

Vehicle searching for hazardous devices or contraband is an expert task which can take several hours. It is best left to skilled vehicle examiners. There are so many places within even a small automobile where a device can be planted that it is difficult to positively clear a car without a detailed mechanical examination. Additionally, an automobile may be booby trapped by many electrical or mechanical systems.

Carefully enforced measures in parking, delivery supervision, etc., as previously discussed, are the keys to anti-vehicle bomb safety.

If a vehicle is suspected of carrying a hazardous device, then an immediate emergency procedure should be enforced. Where vital installations are nearby, or even other vehicles, post-explosion and fire risks are extreme, so maximum precautions must be observed. Vehicle devices require maximum evacuation distances.

When checking a vehicle for a possible bomb, security personnel will require the following information, or as much of it as can be learned:

1. The owner/driver of the vehicle.
2. Is the vehicle stolen?
3. When was it parked?
4. Over what road surface, and for what distance, did the vehicle travel prior to parking?
5. Is it locked, and if so, which doors or windows?
6. When was the car last serviced in a garage?
7. How much fuel is carried, including any spare cans?

Some of this information will not be known, and some will be the task of law enforcement officers called to the scene, who can check current police hot sheets for stolen vehicles. And if there appears to be a strong possibility that the vehicle might contain a device, such areas as doors and windows (item 5) should be checked

BOMB THREAT PROCEDURE

To be used for telephoned / mailed bomb / arson threats:

_____

1. INFORMATION:

Bomb's supposed location ....
....
....

Time of threat call ....

Supposed explosion time ....

Any clues on caller / call ....

....

....

Telephone Operator's ....

TELL OPERATOR TO STAND BY AT THE SWITCHBOARD

2. RISK EVALUATION:

A. How much time is there before the explosion? .... min.

B. Is the device in a densely populated area?   yes/no

C. Does the device endanger critical resources?   yes/no

D. Any reasons to suggest the call is a hoax?   yes/no

E. Could this be a diversion attack?   yes/no

3. INITIAL RESPONSE:

A. If the call suggests a hoax: 1. organise local staff search.
   If manpower/time exists: 2. dispatch guards to area.

B. If call sounds real, risk is high, or little time exists: evacuate personnel to a safe area.

C. If call could be a diversion: increase security on vulnerable personnel / resources.

4. SECONDARY RESPONSE

A. If call is actual
1. seal entry / exits.
2. dispatch guards to control danger zone.
3. alert bomb squad—tel......
4. alert medical services—tel..
5. alert fire service—tel......
6. alert disaster service—tel..
7. alert support services—tel..

B. If insufficient time exists
1. maintain evacuation for 50 min.
2. organise search after 50 min.

C. If staff search work area
1. dispatch guard to supervise.

D. If call is diversion
1. check for intruders.

5. INVESTIGATION REPORT

A. Has the entire area been searched? yes/no

B. Are ANY suspicious objects reported? yes/no

C. If yes, where is device(s) ....

D. Description of suspicious device:

    size ....

    camouflage container ....

    initiator ....

    explosive ....

    other details ....

E. What risk is the device to
1. RESOURCES hi risk / lo risk
2. PERSONNEL hi risk / lo risk

F. Status of evacuees:
1. safe distance / unsafe distance
2. exposed / behind cover
3. restricting emergency access / clear

G. Is further evacuation necessary: yes / no

H. If yes, to what distance ....

6. EMERGENCY RESPONSE:

A. If no device is found   STAND DOWN SERVICES

                                  police tel ....

                                  medical tel ....

                                  fire tel ....

                                  support tel ....

                                  disaster tel................

B. If device is suspected:
1. check evacuee status for distance, cover and access obstruction.
2. request police support tel bomb squad ....
3. re check emergency service status and increase alert status.
4. dispatch escort for support and police service to ensure fast access.

C. Have other devices been searched for?   yes/no

7. CLEARING OPERATION

A. Have suspected devices been deactivated / removed by EOD? yes/no

B. Have secondary searches been made?   yes/no

C. Have risk areas been cleared and declared safe?   yes/no

D. Have all alerted service been stood down?   yes/no

E. CALL ALL PERSONNEL TO CONTROL CONSOLE.

F. CRITIQUE OPERATION AND PREPARE REPORTS, ETC.

....

Figure 11. Security Bomb Threat Procedure Report

only by a trained bomb technician. Doors and windows are frequently used to trigger an explosive device when opened.

Clues to booby traps and other explosive devices are similar to those previously listed for general bomb searching. *Essentially, if there is sufficient reason to believe a vehicle may be hazardous, it should not even be approached.* The police bomb squad should be called for assistance.

SUMMARY

As an aid in reviewing and summarizing the bomb threat security response procedures outlined in this chapter, a detailed, step-by-step analysis of the responsibilities of all personnel involved is presented below, from the moment the initial threat call is received to the "stand down" and after-the-fact review of the operation.

It will be noted from this review that the security response to the bomb threat does not end when the situation has returned to normal. It is the responsibility of Security, working in cooperation with local police where necessary, to conduct a follow-up investigation of the incident, make a report of all the known facts to the organization's administration, and provide a critique of the operation, including the total security response. Such incident review and reporting is an ongoing part of effective security, since it provides the information essential to improving and/or adapting the organization's response to possible future situations.

In addition to the following review of emergency response procedures to the bomb threat, a dramatization of a typical incident and the security response has been included in this manual as Appendix B.

TELEPHONED BOMB THREAT
EMERGENCY RESPONSE PROCEDURE:

Incident Situation	Personnel Command	Personnel Assignment
Telephoned Bomb Threat	Telephone Operator	1. Refer to check list. 2. Record bomb's location. 3. Ask when it will explode. 4. Delay and hold caller. 5. Trace or record call. 6. Inform security section. 7. Record time of call. 8. If no reply from security within 1.5 minutes, call (police) emergency. 9. Remain at the switchboard. 10. Fill out detailed check list.
Bomb Alert Stage C	Security Chief	11. Assess information from telephone operator. 12. Organize immediate general or local evacuation. IF necessary. 13. Seal entry and exit points. 14. Dispatch senior security officer to reported site of bomb. 15. Alert police hazardous devices squad. 16. Alert medical services. 17. Alert fire services. 18. Place on alert, Stage C, local emergency/disaster plans. 19. Evaluate the need to increase security on vital services, personnel or resources.

		20. Radio policy and up-dated information to senior security officer, who now acts as incident commander.
21. Dispatch available security guards to incident commander, as requested, to search and control the area.
Bomb Alert Stage C	Incident Commander	22. Arrive at incident site, observe area for any suspicious persons or objects.
23. Relay observations and position to Control.
24. Evaluate situation and target risks with Security Chief.
25. Possible tactical responses: a. full evacuation b. localized evacuation c. search
26. If evacuation is used, then supervise correct bomb alert exit and roll call procedures.
27. Seal off safe areas and select incident control point.
Bomb Alert Stage B	Incident Commander	28. Detail search team to assemble in incident control point.
29. Detail staff supervisors to assist search.
30. Brief searchers on situations and search safety.
31. Assign search areas to searchers.

		32. Relay up-dated situation report to security chief.
Bomb Alert Stage B Search	Search Team(s)	33. Search assigned areas.

Bomb Alert Stage B Search	Senior security officer supervising searchers	34. Supervise searchers' safety tactics.
		35. Mark and report clear areas.
		36. If a device is found:
		   a. Mark the area
		   b. Note device description
		   c. Note device position

Bomb Alert Stage A Device Suspected	Incident Commander	37. Withdraw search teams.
		38. Assess the possibilities of additional devices.
		39. Relay up-dated situation to Security Chief, a safe radio distance from the suspected device.
		40. Possible tactical actions:
		   a. personal investigation
		   b. intensify evacuation
		   c. call for emergency services

Bomb Alert Stage A Device Suspected	Security Chief	41. Call for Police Bomb Squad. 42. Intensify all emergency service preparedness status. 43. Dispatch mobile patrol unit to main entry point. 44. Disptach mobile patrol to incident site, if necessary

		to control traffic, and help establish evacuation cordons.

Bomb Alert Stage A
 	
Device Suspected
	Incident Commander	45. Re-evaluate evacuation levels and distances.
		46. Establish perimeter cordons and assure free access of emergency vehicles.
		47. Request Control for any special services required.
		48. Inform Security Chief of:
		    a. Device's location
		    b. Description of device
		    c. Possibility of other devices

Bomb Alert Stage A Device Probable	Security Chief	49. Re-evaluate the need for increasing security on vital services, personnel and resources.
Bomb Alert Stage A Device Probable	Entry Guards	50. Check I.D. of responding emergency units — IF criminal infiltration is possible.
Bomb Alert Stage A Device Suspected	Security Chief	51. Inform Incident Commander of current situation.
Bomb Alert Stage A	Mobile Patrol	52. Escort emergency units to the incident site.

Device Suspected		53. Stand by to act as escort or courier.
Bomb Alert Stage A Device Suspected	Incident Commander	54. Brief Police Bomb Squad commander 55. Request policy decisions from Security Chief.
Bomb Alert Stage A Device Positive	Incident Commander	56. Inform Control, from a safe radio distance. 57. Hand over command to Police Bomb Squad commander, for device neutralization.
Bomb Alert Stage A Device Neutralized	Incident Commander	58. Dispatch mobile patrol to escort Bomb Transporter out of the installation. 59. Ascertain that exits are cleared of traffic.
Bomb Alert Stage A Second Search	Security Chief/Incident Commander	60. Re-assemble search teams.
Bomb Alert Stage A Second Search	Incident Commander	61. Re-assemble search teams at incident control point. 62. Assign search areas. 63. Brief searchers in safety tactics.
Bomb Alert Stage A	Senior Security Officer with Search Team	64. Supervise searchers for safety tactics. 65. Mark clear areas.

Second Search		66. Report clear areas to Incident Commander.

Bomb Alert Stage A
 	
Current Situation Tactical Evaluation
	Incident Commander	67. Inform Security Control/ Chief of current tactical situation.
		68. Inspect area with:
		  a. Senior Security Guard
		  b. Bomb Squad Commander
		  c. Building supervisor
		69. Declare area Safe/Unsafe
		70. Inform Control and Security Chief of updated situation report.

Bomb Alert CLEARED Stand down of emergency forces.	Security Control	71. Stand down all emergency services. 72. Stand down all disaster control services. 73. Return security personnel to normal stations. 74. Consult with Bomb Squad commander. 75. Direct follow up investigative procedures in cooperation with local police departments. 76. Organize review and critique of operation with involved security personnel.

Chapter 6
MANAGEMENT BOMB THREAT PROCEDURES

Not every organization has sufficient security staff, or any security staff, for assignment solely to control bomb situations. Many small organizations must rely upon company personnel for anti-bomb security and for emergency responses to bomb threats.

In such situations, the following planning and procedure guides may be used by management in the development of company preventive and emergency plans.

ESTABLISHING POLICY

The object of bombing and bomb threats is damage, harassment, the fermentation of employee fear, death and injury. The countermeasures to each of these factors is the responsibility of company management.

The development and application of the above countermeasures will involve an expenditure of time and money. However, this expenditure is cost-effective. The risks of bombing and arson through incendiary (fire) bombs is quite high, especially in organizations with racial, political, defense or "establishment" business, personnel or connections. More important, for every bomb exploded, there are *at least* 200 telephoned bomb threats.

The effect of a bomb threat will depend upon the type of organization. In a small factory, a bomb threat will involve time wastage, loss of production and staff fear/harassment. In a theater or restaurant, it may necessitate an evacuation of hundreds of people, the refund of admission prices, and a considerable loss of money through immediate effects, and later bad public relations. In a busy shop or department store, it may mean the evacuation of hundreds of customers and staff, with an enormous financial loss, through loss of income, staff time wastage, and bad public relations.

For these reasons, and many others, anti-bomb/bomb threat security is obviously cost-effective. Further, it is clearly the responsibility of company management.

Basic Decisions

What steps may be taken by management to prevent losses, and to protect their staff and assets from bomb attack and threats?

The first step is to develop a corporate policy for preventive security. Prevention is better than cure, and money or time spent upon the prevention of bombing or arson is highly cost-effective.

Preventive security policy may be developed by a management team, an independent security consultant, or with the advice of local police Crime Prevention divisions. In general, a combination of all three approaches will provide the best security.

The approach above will decide the preventive security measures to be adopted. Initial perimeter protection and anti-intruder protection measures may be handed over to an industrial security and alarm company. It is recommended that an independent security consultant be utilized in instructing the chosen alarm company for necessary protection. This will often save expenditure on unnecessary physical protection equipment.

If the advisors recommend a professional security employee be appointed, or a contract guard company be hired to provide services, then it should be firmly established that these personnel are trained and experienced in modern anti-terrorist and crime prevention techniques.

If the company decides to appoint an employee for security duties, or to advertise and employ a person for these duties, then every effort must be made to ensure that these personnel are trained in anti-bomb/bomb threat security. Personnel may attend relevant seminars or guard instruction courses, or follow correspondence courses in essential subjects.

If a contract guard service is hired, then it should be established that the organization is primarily interested in having guards who are skilled in preventive bomb security, and who can competently handle entry control, searching, anti-bomb patrols, dealing with suspect devices, etc. A security consultant may be able to recommend a contract guard service, but if not, then management should ask for the training background of guards to be supplied.

If it is decided that security will be the responsibility of present management or other key employees, then policy decisions must be made by those personnel. As previously mentioned, a security consultant may be called upon to provide basic plans, and this is strongly recommended. However, for those organizations that will prepare their own plans and policy, the following procedures should be of assistance.

Reliance on Government Agencies

An important factor to be considered in the development of policy is the extent to which responsibilities can be placed upon government agencies.

The local or State law enforcement agencies or the military have many operational duties concerned with bombing or bomb threats on company property. However, there are important limitations to their services. Legal and commercial factors control the amount of advice or service a company will receive from government agencies. For example, the decision to evacuate or not to evacuate, following a bomb threat call, will usually be left to the company. In addition, the company may be largely responsible for searching their premises following bomb threats, because the police or other agencies can provide little real tactical assistance in many situations.

The second policy decision should be to establish a division of responsibilities between company personnel and government agencies. Senior officers from local law enforcement agencies or the State Police should be contacted, and the degree of assistance which they can offer in emergencies should be established.

While in contact with local agencies, management should consider whether their company requires planned assistance from emergency or disaster forces. A company manufacturing hazardous chemicals, or with other potential risks to people or the environment, should be prepared to cooperate closely with local agencies to prepare emergency disaster plans.

The third policy decision is the deployment of staff to handle preventive security and emergency control. This is discussed in the following section.

STAFF DEPLOYMENT TO PREVENTIVE AND EMERGENCY SECURITY DUTIES

An organization's management should deploy key personnel to necessary security duties, as defined by the management or security consultant's report. For most organizations this deployment will be as follows.

1. Preventive Security

Good preventive bomb and other security can be achieved by the education and cooperation of all staff, from management to shop floor worker. There are two main preventive bomb security fundamentals: preventing the access of bomb planters, and speedily discovering planted devices or suspicious objects.

Unauthorized access can be prevented by encouraging any staff to question *any* stranger they see on the company premises, if they are unescorted by a staff member. Any strange person should be greeted with a friendly but inquiring "Good Morning. May I help you?" This is polite and helpful if the person is a legitimate visitor, but it will deter an intruder. It is important that this procedure be encouraged, and that staff do apply it to *any stranger*. It is very easy to walk past company employees and sometimes even security guards if you carry a parcel with the company's address and advertising stickers, or if you just walk past staff and guards confidently with a piece of paper.

Secondly, staff should be encouraged to work neatly and tidily, so that strange or out-of-place objects will be quickly noticed around their work areas. This particularly applies to external areas, such as unloading bays, trash or store areas, etc. A neatly kept area will be easy for staff to cast an eye over for devices thrown over a perimeter fence or wall. All staff should be asked to report any suspicious object to management, if seen in a lavatory, locker room or corridor, and, most important, they must be warned not to touch any suspicious object.

A more specific form of preventive security is in "entry control," that is, the entry of persons, goods or vehicles.

The entry of staff and visitors, servicemen, etc., is usually the responsibility of receptionists or secretaries. These staff should be trained to carefully inspect the authority or identification of visitors, and to have them properly sponsored by a senior company employee. If a visitor is allowed into the company premises, then he should be escorted *in* and *out* by a company employee.

At certain times, or in potential risk organizations, it may be advisable to request visitors to open their work boxes, tool cases or brief cases, so that the receptionist may check that there are no explosive devices inside. This will act as a prevention and a good deterrent, especially if a warning sign to this effect is displayed in the entrance foyer.

Another specific preventive security measure is to deploy supervisory staff to regularly patrol the interior and *exterior* of the organization, to speedily detect planted devices. On such patrols, staff should pay particular attention to the places where an outsider could easily place a device. Good examples of these places are exterior walls, just inside or outside; vehicles parked near the company; window ledges; trash cans; entrance foyers and reception rooms; lavatories and locker rooms, etc. On internal anti-bomb patrols, staff should check trash cans, toilet cisterns, in or over light fittings, fire fighting equipment, behind curtains, and other simple, but easily overlooked, areas.

2. Emergency Responses

All sizable companies will need to develop a telephoned bomb threat emergency response. Various staff will need to be deployed to these tasks, from management to service personnel.

For large companies, it is suggested that three personnel groups be deployed. The first of these should direct these groups, and will be mainly responsible for anti-bomb security, in addition to emergency control. The control of emergencies, such as a telephoned bomb threat, should be maintained by an Emergency Officer/Director. The Emergency Officer should have sufficient appointed deputies to cover the company during holidays and sickness, etc.

The Emergency Officer will appoint an Area Emergency Warden, to cover control in specific or vulnerable areas of the company's premises, where they are large enough to require this.

The third group is the Emergency Search Team, who should be responsible for bomb searching (and preventive patrols) following telephoned threats. These personnel should be sufficient to cover all areas of the company's premises, and be chosen from mature staff who are familiar with the normal objects in their work areas.

The Emergency Search Team, who *must be volunteers* because of the risks involved, may be utilized for other emergencies. For example, they can be trained as First Aiders, Disaster Forces for chemical spillage, Rescue Forces for fire or other disasters, etc.

Detailed plans for bomb threat emergencies have already been provided in preceding sections. However, for the benefit of management who do not employ security specialists, the following guides are provided.

Emergency Officer

The Emergency Officer should be responsible for bomb threat emergency control. While the Emergency Officer should apply his control measures within the accepted company emergency plans and policies, he should take direct control of the situation, and should *not* allow his decisions to be influenced by unqualified personnel, even though those personnel may be senior.

The Emergency Officer will be responsible, within corporate policy, for ordering plant shut-downs or evacuation of staff. Under many circumstances this decision will be the Officer's, and will *not* be made by the police. Therefore, the Emergency Officer should be able to evaluate the degree of threat of a call, and have a thorough knowledge of his emergency plan.

The Emergency Officer, or his management superior, should ensure that the company is protected at all operational times. To assist in this an Emergency Officer Rotation Schedule should be maintained. An example of this follows:

Company Emergency Director ....

Phone .... Radio Bleep No. .... Office No. ....

Deputy Emergency Officers
―――――――――――――――――――――

Shift 1 .... Phone ....

Shift 2 .... Phone ....

Shift 3 .... Phone ....

Assistant Emergency Officers
――――――――――――――――――――――

Shift 1 .... Phone ....

Shift 2 .... Phone ....

Shift 3 .... Phone ....

Emergency Command Post ..... Phone ....

Operational from date / /   - to -   / / .

Figure 12. Emergency Officer Rotation Schedule

Chapter 7

LETTER BOMB SECURITY

Explosive devices can easily be concealed within letters and small packages, and these create a security risk quite distinct from the planted bomb or the bomb threat. Such anti-personnel devices as letter bombs have two offensive functions. They may be directed to kill or maim a specific person, or they may be used to scare and harass the general public.

Although mail devices are ideal terrorist weapons, they are used sparingly. They tend to be selected as weapons in isolated, specific attacks or during a brief, intensive campaign. One reason for this is that terrorists recognize that continuous attacks would only result in having the target organizations obtain full protective security screening equipment. By using random attacks, the letter bomber knows that he can cause chaos without having organizations realize his potential risk over the long term.

Target Risk Evaluation

To develop cost-effective anti-mail device security, each organization will have to evaluate its target risk status. Terrorist targets have included ninety-year-old ladies, children, chemical factories and insurance offices—but obviously not in the same order of frequency or probability. Each organization's anti-mail device security must be in proportion to its target potential.

As with other bomb threats, some examples of high-risk targets are political and racial centers, embassies, defense-related industries, police, fire and government offices, and the like.

Because of the specific anti-personnel nature of mail devices, they create a security risk not only at an organization's facility but also at the homes of key personnel. Mail devices sent to an executive's home place him and his family at risk. Therefore, high-risk organizations, or those involved in political or other controversy, should consider the risks to their executive personnel and offer personal mail screening services if appropriate. A number of mail devices have been sent to politicians and other high-risk individual targets, so this hazard must be considered alongside the organization's internal security plans.

DEVELOPING A SECURITY SYSTEM

Bombing, with the fear it generates, is a form of psychological sabotage. Mail devices have a high fear effect because people realize that the small amount of explosive they contain is sufficient only to cause maiming or hand loss in most cases, as opposed to rapid death. This fear is often greater than that caused by large explosive devices.

Developing a security plan to counter mail device risks has two phases. First is staff education in the recognition and safe handling of suspect mail. The second phase is the installation of screening equipment.

Staff Education

Personnel education against mail devices primarily involves the training of mail reception staff and/or secretaries performing this duty. It is also important for executive staff to be trained against mail device hazards, for security both within their working environment and at their homes.

Mail reception staff should be trained to recognize hazardous deliveries by a program of practical recognition demonstrations. Such personnel should also be taught to know the correct actions to take if a suspect device is encountered.

Because a mail device may only be found infrequently, a constant reminder should be provided to staff in the form of posters and warning notices. Ideally, security intake design should be such that deliveries cannot enter the organization without passing through screening equipment. Since this is not always practicable, the alert awareness of mail handling personnel is the first line of defense.

The incoming mail and deliveries addressed to or intended for potential target risk executives should receive careful screening. This applies to routine deliveries, and especially to personal gifts and unsolicited packages.

To ensure that executives are not attacked through their home addresses, they should also be trained in the recognition of potentially dangerous mail devices. With the escalation in recent years of terrorist actions against executives, and the risks of crimino-political assassination attempts, this training is essential for key executives in many organizations which would once have been considered immune from attack.

Detection Devices

It is difficult to construct an explosive mail device which will pass through the rough and tumble of mail services without being spotted at some point. Small packet and package devices are the easiest to construct and stand the greatest chance of arriving undetected. On the other hand, while the cruder devices are easily recognized, a carefully constructed mail device can be almost impossible to screen out by visual examination, so mechanical screening equipment must be utilized to provide complete security.

There is available a variety of modern security screening equipment, including X-ray examiners, fluoroscopes, metal detectors and explosive vapor detectors. Each has advantages and disadvantages in operation, so a potential target organization should carefully analyze its mail deliveries and risk status before purchasing equipment.

One basis upon which suspect items can be tested would be to check them with equipment which is sensitive to minute traces of Nitrogen Oxides. Such equipment is available, although it is expensive, and it provides a satisfactory means of locating the presence of organic nitrate type explosives such as those upon which gelignite materials are based. If, however, other explosive materials are used, or if gelignite material has been carefully sealed into a completely impervious envelope and the outside of this has been thoroughly cleaned, then such a method of detection may fail.

Fixed or mobile metal detectors are low cost instruments which provide basic screening. They should be considered as a first and most basic line of defense against letter bombs. This method of detection relies upon the fact that it is very difficult to make letter bombs which do not incorporate at least some metal components. If a commercial type of detector is used, then this has a metal case; and neither a percussion fuse nor an electrical fuse can be made without some metal content. Moreover, metal detectors can be very simple, very easily operated, and very cheap in price, and they represent no hazard to the user. Unfortunately, they are also very selective in use and will detect any metal item in a package, responding as effectively to a staple, paper clip or file binder as to a bomb component. In addition, while it would not be easy to make a bomb without metal components, it is possible.

At a higher cost, X-ray examiners or fluoroscopes provide the best way to screen mail. In organizations large enough to utilize central office or security services, such equipment costs can be defrayed by sharing screening services.

Various types of X-ray examiners are available. For large volume mail screening these may be fitted to a conveyor system, with examination taking place at a remote console. Screening with X-rays takes only a few seconds and provides positive security against most devices. Any normal package is translucent to X-radiation and a shadowgraph can be seen where denser materials, such as metals, show up much darker than less dense materials such as paper. Moreover, the shape of items of different materials can be clearly seen, so that batteries, wires or springs can be immediately recognized as such. Even if a bomb were to be made without any metal components—for example, with a friction fuse and a plastic encased detonator—the shape of these and the shape of the explosive charge would be visible in the X-ray image. Even if they were not able to be positively identified for what they were, the suspicious nature of the package could be confirmed sufficiently for it to be referred to the attention of an expert.

To examine larger deliveries, the metal detector may be used in a hand transportable mode. Additionally, mobile X-ray examiners may be used. Some portable configurations may be moduly fitted into inspection consoles and used to examine incoming hand baggage, attache cases, etc., as well as mail.

The radiation emission levels of most X-ray examiners are very low, because they utilize image intensifiers to provide penetrative inspection. Virtually no special design modifications or safety measures need be adopted for operating staff.

Explosive vapor detectors provide effective mail device screening, although they take longer to operate. Adapted to a mobile role, vapor detectors are multi-purpose instruments that may be used to screen incoming personnel, baggage, vehicles and large deliveries.

RECOGNITION OF HAZARDOUS MAIL DEVICES

Explosive devices, incendiaries and booby traps may be constructed to fit within almost any familiar object. The list includes all types of mail deliveries. Such devices are usually constructed so as to detonate as the package or letter is opened. The method of explosive activation can be by the release of a spring, a spring-loaded striker system, or a make/break electrical circuit.

A bomb may easily be constructed in a parcel, which can, of course, carry large amounts of explosive. However, the most common mail devices encountered have been sent in large envelopes or small packages. The typical letter bombs used up to this time have fitted into envelopes roughly $5\frac{3}{4}''$ x $4''$ x $\frac{1}{4}''$ and weighed between two and three ounces.

Detection of Letter Bombs

In order to function, a letter bomb must contain:
1) Explosive
2) A detonator
3) A fuse to fire the detonator.

The explosive is usually a "fast" explosive of the gelignite class, since this will explode effectively without being in a bursting case. Other explosives, including certain gunpowder type mixtures, can also be used.

The detonator will commonly be of the kind which is used with mining and quarrying explosives. This is usually cylindrical in form,

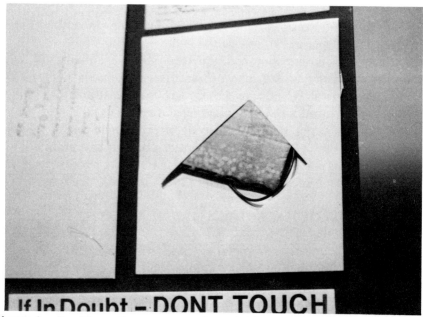

Letter bomb with portion cut away to reveal inner enclosure, sealed down, and electrical wires.

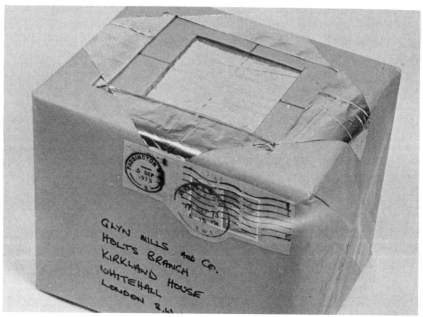

Parcel bomb which detonates when a knife connects the two layers of tin foil. Such a booby trap device shows the importance of never touching a suspicious object.

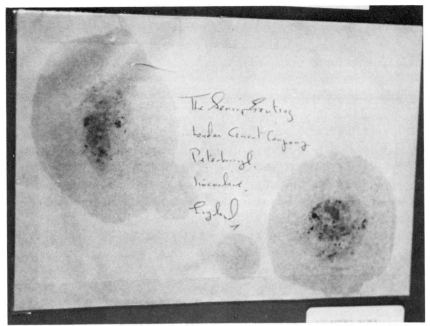

Example of a letter bomb. Greasy patches on envelope and tape all around edges are clues to concealed explosives.

Letter bomb of Mid-East origin. Notice Arabic style of handwriting; sweating explosive stains.

has a thin copper casing and is about 3/16" (5mm.) diameter.

The fuse will normally fall into one of three classes:
1) Friction
2) Percussion
3) Electric.

A friction fuse works rather in the manner of a match head on sandpaper. It is very simple but has a rather high failure rate.

A percussion fuse operates like a firing pin in a gun. A spring is released to carry a striker to impact upon the detonator.

An electrical fuse must have batteries and wiring connecting these to the detonator with some form of contact mechanism to close the electric circuit when the bomb is to be fired.

It is useful to consider these basic essential features of letter bomb construction in looking at the most satisfactory manner of detecting bombs.

The first point that can immediately be seen is that a letter bomb cannot be made of extremely small size or very thin. It could not, for example, be made the size of an ordinary envelope containing an ordinary single sheet of letter heading, or a single order or invoice. This is of considerable importance, since it means that all mail falling within this classification of size and thickness does not need to be inspected. For the majority of recipients, this will mean that more than half of the number of items which are received do not have to be inspected.

Where larger packets are considered, it will be apparent that, at least from the point of view of size, it is possible that they could contain a bomb. If, however, they carry, on the label or on the outside of the package, the name and address of the sender, and if this is known to the recipient, then it is probably quite in order to regard them as being safe.

If mail is intelligently inspected and these two factors are taken into account—the size of the letter or package and familiarity of sender identification—it will be seen that the number of potential suspect items which need to be inspected is only a small proportion of the total that are received.

Recognition Clues

Some mail devices may be recognized by visual inspection. Mail handling personnel should be alert to a number of recognition clues, the most common of which are detailed below.

1. *Place of Origin.* Note the delivery's post mark. This may be from a country directing a terrorist campaign or from an area where such mail devices are popularly associated. If the arrival of such mail is uncommon, it should be treated as suspect.
2. *Sender's Writing.* Any mail should be treated with caution if it features a foreign style of writing, not normally received, on the address. This should be considered in relation to No. 1, above.
3. *Balance.* Any letter should be treated as suspect if it is unbalanced, has loose contents, or is heavier on one side than the other.
4. *Weight.* If a package or letter seems excessively heavy for its volume, it should be treated as suspect.
5. *Feel.* If an envelope has any feeling of springiness at the top, bottom or sides, but it does not bend or flex, this is a key sign of an explosive device. WARNING — EXAMINE MAIL GENTLY!
6. *Protruding Wires.* Mail devices are often loosened or damaged by rough mail service handling. It is possible that fuse or electrical wires and connections will become loose and penetrate the wrapping or envelope. Any such device is unstable and highly dangerous. It must not be touched.
7. *Holes in the Envelope or Wrapping.* An explosive mail device which has been handled roughly may show wire or spring holes in its outer wrapping. This, by itself or in combination with the other clues described, should alert mail handlers to a suspect device.
8. *Grease Marks.* Certain types of explosives leave greasy black marks on paper, a certain clue to the presence of a suspect device. It can also mean that the explosive has become old and unstable, making it very dangerous.
9. *Smell.* A smell suggestive of almonds or marzipan, or any other strange smell, is a good clue to a suspect device.
10. *Unrequested Deliveries.* Unrequested deliveries, especially packages, should be screened and treated with caution. A

An ordinary small tape cassette box. . . .

. . . opens to reveal an explosive/incendiary device.

book or other thick brochure discovered upon opening a delivery should be reported to the security department or examined for any of the above clues. Any mail over which there is the slightest suspicion should *not be handled.* Remember the catch phrase—TOUCHING TRIGGERS TRAGEDY.

(It is of benefit to both parties if senders place their name, organization address and telephone number on packets. Then, in cases of suspicion, they may be contacted for an explanation of the contents.)

11. *Suspicious Packaging.* If an envelope is taped down all around, instead of having a normal opening flap, it may contain a booby trap spring. Such letters should be handled very carefully and treated as suspect.
12. *Letter Stiffness.* Feeling *very gently* should reveal whether an envelope contains folded paper or a device. The presence of stiff cardboard, metal or plastic should alert the handler to a possible suspect device.
13. *Inner Enclosures.* If, after opening a letter or package, the mail handler encounters an inner sealed enclosure—whether or not it fits any of the above descriptions—the item should be treated as suspect.

SECURITY RESPONSE TO SUSPECT HAZARDOUS MAIL DEVICES

Letter and package devices usually contain only a small amount of explosive. Their intended effect is to injure, maim or even kill the person opening the device. Thus, if a letter bomb is suspected, the first rule is not to allow anyone to open or handle it.

As in other bomb situations, one should never assume that only one device exists. If a suspect letter bomb is found, then the remainder of the mail—and all deliveries over the next few weeks, at least—should be carefully examined.

It is reasonable and expedient to assume that any mail device which has arrived safely by post will not explode unless grossly mishandled. However, if a device has become damaged in transit—a very real possibility—it may be highly sensitive and unstable.

The following security procedures should be followed if a

device is found that exhibits one or more of the warning clues detailed above.

A. *If the suspect device has not been touched*
 1. If a suspect delivery is spotted, do not allow anyone to touch it.
 2. Evacuate the room. If the device appears to be very large, surrounding rooms should also be evacuated.
 3. During evacuation, leave doors and windows open, so as to reduce any blast effects.
 4. Keep people away from the area.
 5. CALL SECURITY OR POLICE EMERGENCY.
 6. Do *not* handle the suspect object—and do not try to carry it outside.
 7. Do *not* place the device in water.

B. *If an item is suspected during handling*
 1. Place the suspect item in a corner of the room, handling it very gently and making sure not to turn it over or unbalance it.
 2. Make sure that the device is placed away from windows, and that the windows are open.
 3. Evacuate the room, and surrounding rooms if necessary. During evacuation, leave doors and windows open.
 4. If possible, place the device in a bomb bunker or under a blast suppression blanket. (TO BE ATTEMPTED BY TRAINED SECURITY PERSONNEL ONLY.)
 5. Keep people away from the area.
 6. CALL SECURITY OR POLICE EMERGENCY.
 7. Do not try to carry the device outside. Use procedures 1 thru 6 as above.

A number of key safety principles are embodied in the above listing of procedures. The first is that suspicious mail should not be handled any more than absolutely necessary, and then only with great care. No attempt should be made to open or look inside of suspicious packets or letters. Corollary advice is not to place a suspect device in water or attempt to carry it outside. Since the application of any weight or pressure might trigger the explosion,

LETTER AND PARCEL BOMB GUIDE

Recognition:

Suspect a letter / parcel bomb if —

- It has a foreign address or style of writing.
- It is unbalanced.
- It feels springy (handle very gently).
- Small wires poke out.
- There are pin pricks or holes in the envelope.
- It has greasy patches.
- It has a strange smell.
- It is taped or sealed down all around.
- It feels too stiff.
- It has an inner sealed enclosure.

What to do:

1. Place the suspect device in a corner of the room, away from windows. Handle it *gently* and do not turn it over.
2. Evacuate the room, and surrounding areas if necessary.
3. During evacuation, leave doors and windows open.
4. Keep people away from the area.
5. Call SECURITY on Ext. _____.
6. If no reply, call POLICE EMERGENCY.

If you receive an unexpected or unrequested delivery, phone the senders for an explanation of the contents.

IF IN DOUBT — DON'T TOUCH!

Figure 13. Letter and parcel bomb guide

suspicious deliveries should not be covered with sandbags or *any* other objects. (The exception, as noted, involves the use of bomb buckets or blast suppression blankets by trained security personnel.)

In climates where letter bombs are a continual hazard, the risks must be continually countered in order to maintain the safety of personnel. Posters and displays can be used, as previously mentioned, to provide emergency instructions and basic recognition advice about letter bombs. Figure 13 is an example of such a notice; it can also serve as a summary of the key elements in letter bomb security.

Appendix A

BOOBY TRAPS, MINES AND IMPROVISED EXPLOSIVE DEVICES

Booby traps are anti-personnel devices, used tactically as a public terror weapon, or to strike specific strategic or tactical targets. Booby traps or heavily disguised explosive devices have an effect which is far greater than their physical blast and casualties. In a bombing campaign, whose main objective is to reduce public morale and confidence in the government or to create financial de-stabilization, an occasional booby trap is important in maintaining the impetus and state of fear of the campaign. An occasional booby trap, or disguised device detonated in a public place, will produce weeks of wasted time and effort by security forces and public police, through the investigation of hundreds of false alarm "suspicious object" calls from the public.

Perpetrator Motivation

The perpetrators of booby trap and improvised explosive device incidents come from a varied group. This includes Special Forces of military units, urban and rural guerrillas, terrorists and crimino-terrorists, criminals (organized crime syndicates), psychotics or pyromaniacs, and schoolboys.

This broad group of perpetrators, which encompasses widely differing motivations and levels of technical skill, may seek equally varied goals. Strategic attacks are made to kill and maim, or cause materiél damage, intended to lower public morale and create political pressure or ill feeling between political or racial groups. Other motivations include personal or financial gain, psychological satisfaction or a desire to experiment.

Target Selection

These varied motivations will decide the target selected by the perpetrator(s). The target may be a "direct" attack, by sabotage of government property or personnel. It may, however, be "indirectly" aimed at the government, through reducing public confidence. Indirect attacks are mainly used to reduce morale and exert political pressure, without alienating the perpetrators' cause from the general public.

Probable and in-past targets are:

1. Government/Military installations, vehicles, equipment and personnel.
2. Public transportation, road, air and sea.
3. Public gathering places, transportation centers, entertainment industries, meeting or demonstration places, tourist or commercial zones, etc.
4. Embassies and Consular offices, vehicles or personnel, at their official or private residences.
5. Public Utility Services, mail, power, communications, water and sewage, etc.
6. Industry, especially defense or government-related industries.

TACTICAL AIMS OF BOOBY TRAPS & I.E.D.'S

Booby traps and I.E.D.'s may be divided into two main tactical divisions: those intended to harass the general public, and those intended to harass security forces. Within these two tactical parameters, there are five basic types of booby traps or I.E.D.'s:

1. Delay-activated devices, camouflaged in packaging or containers.
2. Anti-handling devices disguised as common objects.
3. Anti-handling devices camouflaged in packaging or containers.

4. Devices operated by man or vehicle movement.
5. Devices operated by direct or remote control.

With these five basic types of traps and devices, low cunning and infinite variety are easy to apply to construction. To see how cunning and varied traps and devices can be, let us examine these five basic types in more detail:

1. Delay-Activated Devices, Camouflaged in Packaging or Containers

Mainly used to strike against the general public, when left in a public or well-frequented place, these devices are usually hidden in a container, such as an attache case or shopping bag, or packaging like a parcel or box. These devices usually have time-delay activators, designed to detonate the device within minutes or hours of planting.

2. Anti-Handling Devices, Camouflaged in Packaging or Containers

This type of device is usually of fairly simple construction, and can easily be de-activated by interrupting the firing train. However, the device may incorporate an anti-handling actuator, designed to detonate the device when it is handled—the tactical aim being the harassment of the general public or responding security forces. The anti-handling actuator can work on several different systems. First, the approach to the device may be mined or trapped with trip-wire-actuated explosives. Secondly, the device or its container may be booby trapped to actuate its device when it is touched, approached or moved. Third is the most dangerous form of booby trap, which activates when an Explosive Ordnance Disposal operative attempts disarming or neutralizing the device.

3. Anti-Handling Devices Disguised as Common Objects

ANY OBJECT MAY BE BOOBY TRAPPED. The art of booby trapping is "infinite variety and low cunning." Variety is effective in booby trapping by trapping widely differing types of objects. Low cunning is effective by trapping the types of objects which people would never expect to be dangerous. Just a few objects which can be trapped are dead or injured bodies, excreta, contraceptives, sanitary towels, letters, books, light bulbs, etc. The favorite and most effective objects to booby trap are those which are quickly or unconsciously handled, such as arms finds, pornographic books or souvenirs, light switches or doorhandles.

4. Devices Operated by Man or Vehicle Movement

Devices activated by man or vehicle movement are commonly used. This type of device may be as simple as a mine placed under a road, or a trip wire activating a device beside the road. However, it could be as complex as a device which activates when an illegal poster is pulled off the wall, or when a man walks through an invisible light beam.

5. Devices Manually Operated, or Operated by Remote Control

A device may be activated by direct control with a pull wire, or by remote radio control. These types of devices are usually used to harass security forces who are on patrol or have been lured into the killing zone of a device or staged incident.

These, then, are the five basic types of booby traps and I.E.D.'s. They may be constructed from a variety of commercial, military or improvised components.

THE CONSTRUCTION OF BOOBY TRAPS, MINES AND I.E.D.'S

Military and commercial explosives and device components are often stolen internally and externally, and these frequently form part of improvised hazardous devices. Both simple and sophisticated devices may be improvised from easily and legally purchased items.

Some of the sources for improvised device components include drugstores, chemical factories and stores, hardware stores, electronic and model suppliers, construction, demolition and quarry/mining sites, military bases and field firing ranges, machine shops and garages, school, hospital and research laboratories, paint stores, agricultural chemical stores, firework and small arms ammunition, etc.

Although booby traps and I.E.D.'s may appear in an infinite variety of forms or disguises, the technical requirements for a "firing train" necessitate a certain number and order of common components. These common components are igniters, switches and initiators, fuses, booster and main explosive charges.

We can divide the above components into three operationally constructed types: military, commercial and improvised.

Military Components

Military based devices are activated by switches or initiators which have been stolen from military stores. These include pressure switches, release-of-pressure switches, pull switches and chemical delays, lead delays, clock timers, safety fuses and various detonators. Because of their high safety factors these components make the most reliable devices. They are used with standard military explosives or propellants, such as primers, wet gun cotton slabs, CE/TNT charges, plastic explosive, sheet explosive, grenades/rockets, shells, small arms ammunition, etc.

Commercial Components

Commercial components are usually stolen from mining, quarrying and demolition sites. These are frequently raided by criminoterrorist organizations and others. Commercial activators are mainly manually or electrically operated switches, fog detonators, electric or pyrotechnic detonators and blasting caps, etc. These are used to activate explosives such as gelignite, dynamite, Nobels 808, etc.

Improvised Components

Purely improvised devices are most variable in their construction and components. They may include any of the components described above, or be solely constructed from improvised components. Improvised devices will be constructed from the following components:

Igniters:
1. Thermal: These include steel wool and batteries, fuse wire,

2. Chemical: heating filaments, hot wire bridges, light bulbs filled with black or photoflash powders, gasoline-filled light bulbs and model aircraft glo-plugs.
Including weedkillers and sugar (also used as an explosive), chemicals which ignite when saturated with sulphuric acid and numerous other chemicals which ignite after contact.
3. Propellants: These may be used as percussion type igniters, or when converted to a powder train fuse. Rifle, pistol and shotgun cartridges may all be used as igniters, especially in pressure and pull (trip wire) devices.
4. Frictional: Several chemicals will ignite with friction. Two common examples of these are Mercuric fulminate and Nitro iodide. A hard blow, friction or thermal action will ignite these highly sensitive chemicals. Other improvised frictional igniters are match heads and similar substances.

Fuses:

Various pyrotechnic fuses may easily be improvised by a bomber. There are four basic types of improvised fuse: firework fuses, cigarettes, straws and impregnated string.

Firework fuses or complete fireworks may be used in improvised devices. These can be obtained seasonally and stock piled, or obtained from theatrical suppliers. Many fireworks contain a small firing train, and thus provide a complete activation system for explosives.

Cigarettes are a reliable slow burning fuse, which can be attached to a variety of pyrotechnic or explosive devices.

Drinking straws make ideal improvised fuses, which may be filled with various slow burning chemical mixtures, such as Potassium permanganate, flour and sulphur.

Finally, cotton or rope/string may be impregnated with various inflammable mixtures, such as glue, black powder, sodium chloride and sugar (explosive) and potassium chlorate and sugar (explosive).

Explosives:

Explosives may be improvised from dozens of common household, medical, agricultural and industrial chemicals. The list is too large to include. The best advice is to treat *any* unknown liquid, solid, powder or crystals as dangerous.

Activator Switches:
A booby trap or I.E.D. may be activated by at least twenty different methods, or combinations of these methods. Activation methods can be divided into the tactical aims of devices, which we described earlier.

1. Simple delays designed to detonate a device at a predetermined future time.
2. Methods designed to kill a target moving in the device's kill zone.
3. Methods designed to kill a target handling the device.
4. Methods designed to kill EOD operatives attempting neutralization of the device.

The switch method may be specific for one device, or have all four incorporated into the device. Therefore it could be constructed so as to activate when touched, when a person enters its kill zone, if an EOD operative attempts neutralization, or by simple time delay.

The exact type of switch activator will depend upon the four tactical aims described above. These are:

1. Delay Activators
A device's explosion may be delayed by using a candle or cigarette and improvised fuse as a delay. These are the simplest types of delay, but more complex versions include acid type delays, pyrotechnic chemical delays, metal fatigue (stretched solder wire), watch and clock mechanisms, slow moving parts of machinery (sabotage technique), water reactions and expanding de-hydrated foodstuffs (e.g., rice devices), and decaying circuit from an aging battery, plus the actions of solar heat and oven heat.

2. Handling Activators
A device may be planted so as to kill a person who handles it. The usual activators in these devices include commercial or improvised trembler contact, rolling ball bearing switches,

inertia, magnetic and spring devices constructed from clothes pegs, mousetraps, knife and hacksaw blades.

3. Activation by Target Movement in Kill Zone

Two activation methods are used. First, direct control from a pull wire, distant electrical contact, and by remote radio control, using a small radio transmitter. Secondly, using increase/release of tension trip wires, pressure switches or photo-electric beams. These activation types are often used to kill EOD personnel en route to or attempting neutralization of the device.

4. Anti-Handling Activator

Almost any conceivable system may be used to construct an anti-handling device. First, it will probably be so simple and obvious that EOD personnel will casually overlook it. Second, it will be highly complex and so well disguised that EOD personnel fail to spot it. Activators and switches used for anti-handling devices include ALL of the methods we have previously mentioned for activating devices AND MANY MORE. Further, as with many devices, more than one activation system may be incorporated in one device. Indeed, with a complex device system, the approach route may be mined or, connected to a trip wire activator, it may be directly or remotely controlled by an observer watching the EOD team, and it may incorporate several passive anti-handling activators internally.

5. Advanced Activators

Activation systems are by no means limited to the many examples which we have outlined above. Complex switches from barometers and altimeters may be used, probably to detonate devices in aircraft or sub-surface naval craft, designed to activate at a predetermined altitude. Wind-generated switches are easily improvised, or a radio activator made to detonate a device when security forces use their radio frequency near a device. Further examples include acoustic activators, radiographic sensitive activators (designed to detonate when a device is X-rayed). These latter complex systems have NOT YET been used operationally.

DETECTION AND AVOIDANCE OF BOOBY TRAPS, MINES AND I.E.D.'S

The best advice we can give for detecting and avoiding devices is to *trust nothing and assume that nothing is safe.*

We have shown that booby trapping may be accomplished with a wide variety of improvised components and devices, and further, that these may be activated by over twenty different systems, disguised in any common object. However, working within these wide parameters, it is possible to provide some specific advice on the detection of "probable" and previously used devices—but always bearing in mind the risks of the "improbable," which make the best booby traps.

We can first divide booby traps, mines and I.E.D.'s into two operational classes, EXTERNAL and INTERNAL. Let us now examine each of these classes, and identify the probable high-risk target areas.

External

Devices planted in an open area, whether a forest, garden or city street, must come into contact with their targets. First, trappers place their devices where vehicles or personnel will pass near them. Secondly, they place a trap so as to "lure" personnel within their killing zone.

The first category features devices which are placed in areas of known personnel or vehicle movement. The most obvious examples of these sites are frequented roads or paths, railway lines, airport runways and shipping routes, etc. These places may be trapped with pressure-sensitive devices, or may be crossed with pull/release tension wires, magnetic devices, etc.

If a bomber's target is trained or experienced personnel, then he must allow for their trained trap avoidance tactics. He will create a diversion by placing trees or abandoned/accident/crash vehicles in the road. Then, the obvious avoidance detour route will be mined or trapped. However, a clever bomber will use a double bluff against experienced personnel. He will create a route diversion, as we outlined above, possibly using a realistic looking road accident, but the first, most obvious detour will not be trapped. Instead, a bluff is used, for the trapper places devices on the secondary route. These bluff tactics can be used on a major street or a forest track, where the diversion is a fallen tree or bushes.

As we pointed out earlier, booby traps are particularly effective when they work on instinctive human actions. Thus, a release-of-tension or increase-of-tension device which activates when an annoying tree branch is swept away from a forest patch is highly effective. The trapper can often take advantage of trained security force personnel's reactions. He may, for example, halt a vehicle convoy or foot patrol by a diversion, explosion or small arms fire. He will previously have planted devices in the areas of cover and deployment taken by the force's personnel. Against inexperienced personnel, he may use a simple device system, where explosives are placed on a roadside ditch or hedge. However, he may use a bluff against experienced personnel, and leave these areas clear. Instead, he mines the areas slightly beyond, i.e., the personnel's second choice of cover, which they will presume is safe.

The probable place to find these types of simple or bluff devices are cuttings, embankments, blind bends, bridges, culverts, obstacles and the areas around them (bluffs), posts, hedges, wooded stretches, road junctions and cross roads, etc. These locations should be treated as hazardous, especially when they are on the route to incident sites or after attacks at vulnerable installations. They are potential sites for hazardous devices, sniping and ambush.

Our second type of device is planted so as to "lure" personnel into its killing zone. These devices usually feature a "lure" and camouflage. The "lure" can be an "unexploded" device (which probably has an anti-handling device incorporated), an arms cache, dead or injured bodies (enemy or friendly), excreta, or "enemy" clues, such as clothing, weapons, food or souvenir items. The device itself may be trapped against proximity movement of vehicles or personnel, or activated by handling. It may, however, be controlled directly/remotely by a distant observer, who is using a pull wire or radio transmitter.

Internal

Because of the greater number of objects which may be trapped or mined, there are more possibilities for booby trapping on the interior of a building. The types of trap will probably fall into four main sections:

1. Devices planted and operated by moving a door, window or machinery.
2. Devices concealed and activated inside equipment and fittings.
3. Devices concealed in the structure.
4. Planted objects.

Let's examine these in order:

1. Internal Devices Operated by Movement

Any object which may be moved is vulnerable to booby trapping. The movements of an opening door, window, door handle or garden gate are ideal for increase/release of tension-activated devices, electrical contact and mechanical activator systems. If the bomber's aim is sabotage or long term harassment, then he may booby trap machinery, utilizing the delay of parts which gradually come into contact and activate an electric device.

House garden or factory/military base gates, entrance doors, windows, attic doorways are all prone to this type of trapping and should be approached cautiously. Other items which make effective nuisance targets are moving machinery on which electrical contacts can be placed, such as lathes, cranes, fork lift trucks and drills, etc. Entrances will probably be trapped with devices activated by movement, such as increase/release tension wires and pressure switches, plus make/break electrical circuits. Machinery will be trapped with electrical contact make/break circuits primarily, usually featuring contacts on either side of the two moving parts, say a drill head and bit contacting the base plate.

2. Concealed Devices Activated Inside Equipment or Fittings

These are very effective devices, because they are camouflaged in common containers and are often handled unconsciously. An excellent example is a portable radio, tape recorder, television or washing machine, packed with explosive and electrically activated when switched on. A similar example is a light bulb filled with explosive or pyrotechnic compound, again electrically initiated when switched on. These are simple traps, but ones which are highly effective because they make use of everyday unconscious actions.

The main objects to suspect are any electrical fittings, radios, tape recorders, walkie-talkies, record players, light fittings and similar items which have electrical switches, and which may be wired up to electrical detonators or heat-sensitive chemicals. Gas ovens and gas bottles are subject to booby trapping, and their content itself is highly explosive.

Secondary objects to suspect are cupboards, drawers, filing cabinets, refrigerators, cocktail and drink cabinets, etc. The doors of these objects may be booby trapped, and activated whether they are opened or closed. A good trap is to leave a drawer open a little, with an electrically activated device which works whether a person opens or shuts the drawer. This is a good device because it works on two unconscious human reactions: first, the "tidyminded" person who naturally shuts the drawer, and secondly the "curious" who opens it to look inside.

3. *Devices Concealed in the Structure*

A device may be concealed in a building's structure by simply shoving it under a loose floorboard or behind a seat cushion. It could, however, be carefully hidden inside a hollowed out wall or pillar, and covered with plaster or cement. Such a device was used during World War II in an attempt to assassinate Führer Hitler. A device was built into a pillar several months before Hitler was to occupy a building.

Devices may be concealed in the structure and activated by long time delays or remote control, although this type is not very common. More common are types which are activated by increase/release of tension and pressure switches. The floors may be mined by increase-of-pressure switches. These usually activate when trodden on; however, they can activate either when trodden on or when an object is lifted off them. This second type of device is used on dead bodies and arms or explosives caches, plus common furniture. Increase- and release-of-tension devices are connected to actuating trip or pull wires, stretched across a room, corridor or stairs, or attached to movable objects. These wires may be connected to devices hidden in the structure, possibly some distance away.

4. *Planted Objects*

Our last category is planted objects. These are usually small devices designed to harass and delay security forces. Two basic types are used. The first is intended to harass immediately responding security forces, and the second is intended to harass the general public or security forces over a long period of time.

The first group is intended to harass security forces and EOD personnel. These tend to be small objects which will be quickly or unconsciously handled, such as evidence, pornographic books, anything likely to be picked up as a souvenir, cooking utensils, food or illegal/underground literature.

The second group, intended to harass over a much longer period, is restricted to items which will not be handled too quickly, and preferably not until the building is "searched" and "cleared." These devices are usually concealed in books on shelves, bottles, inside food packages or containers, clothing and other items which may be overlooked in a brief search.

CLUES TO BOOBY TRAPS, MINES AND I.E.D.'S

So far we have outlined the probable places for devices, their commercial, military or improvised components, and the tactical aims of various devices. If personnel are alerted to the risks of devices in these locations, and can recognize their components, camouflage or disguises, then they will readily detect clues to their presence. Unfortunately, clues may not be present. A neat and tidy bomber will leave no trace of his work, unless it is a "lure" to draw EOD personnel into a kill zone. However, an untidy or ill-trained bomber may leave visible clues, which include:

External Clues

    Abandoned vehicles
    Abandoned vehicles on routes
    Accidents/crashes on routes
    Disturbed ground, small hollows after rain
    Military containers of ammunition/explosives
    Footprints or vehicle tracks for no apparent reason
    Vegetation camouflage, cut, bent or withered
    Cut vegetation
    Heaps of leaves or scrub
    Marks on trees (indicating traps to the enemy)

Internal Clues

Obviously out of place or "lure" objects
Explosive wrappings, ammunition or explosive containers
Sawdust, brickdust or metal filings
Dusty footprints
Scratched or new paint and timber
Clothes pegs, nails, electric leads, tin foil, string, wire, mousetraps, watch and clock parts
Marks on walls (indicating traps to the enemy)
Loose floor boards or raised carpeting, tiles, etc.
New brickwork, plastering or concrete

Appendix B

DRAMATIZED BOMB THREAT SECURITY RESPONSE

The following dramatization, describing a bomb threat incident in a large computer installation, has been created to illustrate the emergency response to a telephoned bomb threat. While all of the procedures included in this hypothetical situation will not be required in every instance, as many as possible of the procedures outlined in Chapter 5 have been included here to exemplify the telephoned bomb threat response.

The Security Department of InterComputer, Inc., is manned by the Console Operator and Security Chief. They have a large force of mobile and other guards who are involved with the operation.

It is late afternoon of a regular working day when a call is received at the main switchboard.

SWITCHBOARD OPERATOR — MARY ADLER
16:15:00   Good afternoon, Intercomputer, Inc. Can I help you?

VOICE ON PHONE
Listen, there's a bomb in your computer room going off in twenty minutes.

MARY ADLER
Hello... hello, I'm sorry, this is a very bad line. I

can hardly hear you. Could you repeat that, please?

VOICE ON PHONE
Don't mess me up. There's a bomb in your computer room. You've got twenty minutes.

MARY ADLER
Hello — don't hang up. Are you serious?

She is answered by silence. Realizing that the caller has hung up, the operator glances at her bomb threat check list to verify the number for Security.

MARY ADLER
16:16:00 Hello, Security Department? Mary here at the switchboard. I've just had a bomb threat. Yes, just now. He said it was in the computer room and going off in twenty minutes.

Her call is answered by Security Control Operator Mike Jones.

MIKE JONES — CONTROL
16:16:05 Okay, Mary, don't worry, I'll handle it. Stay where you are and I'll send a guard over to get more details. (PAUSE) Oh, Mary — keep this under your hat. We don't want to scare the staff unnecessarily.

The conversation has been overheard by John Williams, the Security Chief.

WILLIAMS — SECURITY CHIEF
What's up, Mike?

MIKE JONES — CONTROL
16:16:15 Bomb threat, Sir. Supposed to be in the computer room and exploding in twenty minutes — that's 16:36, Sir. Mary Adler took the call, so I'll get a guard over there to see what clues we can get on the caller.

WILLIAMS — SECURITY CHIEF
Okay, get to it. Who's the senior mobile patrol guard on duty?

MIKE JONES — CONTROL
That would be Sergeant Connors.

WILLIAMS — SECURITY CHIEF
Send him over to check out the computer room and act as Incident Commander. We'll size up the situation before we commit ourselves.

MIKE JONES — CONTROL
16:16:30 Control to Unit 5.

UNIT 5
Unit 5 receiving. Go ahead, Control.

MIKE JONES — CONTROL
Unit 5, we're under a bomb threat. Go to switchboard and question operator Adler. Investigate and report. Over.

UNIT 5
Copy, Control. Ten-four.

MIKE JONES — CONTROL
16:17:00 Control to Mobile 7.

MOBILE 7 — SGT. CONNORS
Mobile 7 receiving. Go ahead, Control.

MIKE JONES — CONTROL
16:17:10 Go to the computer block and stand by until advised. Act as Incident Commander on bomb threat in computer room. Report arrival. Control, out.

MOBILE 7 — CONNORS
16:17:14 Copy, Control. Ten-four, on my way.

WILLIAMS — SECURITY CHIEF
Okay, Mike, we'd better move fast. I had word from Don Riley at PD HQ that there's a radical group after our type of industry. This could be a real bomb.

MIKE JONES — CONTROL
We're a pretty vulnerable target right now, Sir. We've got those government missile programs running on the computer, and the Air Force General's visit.

WILLIAMS — SECURITY CHIEF
16:17:30 Damn, the chemical stores with those flammable liquids are right next door to the computer block. We can't afford an explosion with 300 staff in the danger area. (PAUSE) Right, let's put the emergency plan into action. Mike, you take the radio and I'll handle local coordination. Let's get moving.

MIKE JONES — CONTROL
16:17:45 Okay, Sir, we've got eighteen minutes to go.

UNIT 5
16:17:55 Unit 5 to Control.

MIKE JONES — CONTROL
Go ahead, Five.

UNIT 5
I've checked out the bomb threat call, made from a local pay phone by a young male. Device supposed to be in main computer room and detonating at 16:36. Over.

MIKE JONES — CONTROL
Copy, Unit 5. Ten-five at that ten-nine. Control out.

UNIT 5
Ten-four, Control. Unit 5 out.

MIKE JONES — CONTROL
Control to Mobile 7.

MOBILE 7 — CONNORS
Mobile 7 receiving.

MIKE JONES — CONTROL
16:18:30 Mobile 7, your ten-nine, please?

MOBILE 7 — CONNORS
Mobile 7, I'm in the computer block car park. Over.

MIKE JONES — CONTROL
Your ten-thirty-two report? Over.

MOBILE 7 — CONNORS
No suspicious objects or persons visible. Over.

MIKE JONES — CONTROL
16:18:45 Okay, Mobile 7. Be advised that bomb threat alert may be actual. Stand by to evacuate area. I'll send back-up. Over.

MOBILE 7 — CONNORS
Copy, Control. Ten-four, out.

Security Chief John Williams, handling local telephone coordination, is on the phone with the first of his calls.

WILLIAMS — SECURITY CHIEF
16:19:00 Hello, Fire Department? Emergency, please. Hello? Yes, this is InterComputer Security service. We've got a bomb threat on here, close by our flammable chemical stores. Could you go onto stand-by? Okay, I'll get back to you — thanks.

He breaks the connection and quickly dials another number.

> Police Department? Emergency, please. Hi, this is John Williams, InterComputer Security. We've got a bomb threat on our computer room and chemical stores. We're a vulnerable target right now. Can you put the Bomb Squad on alert? Thanks, I'll call you the second we have anything definite.

Williams then completes his third call, this one in-company.

> Hello, Health Service? Hello, Sister Daniels, this is John Williams at Security. Look, keep this under your hat but I've got a bomb scare on. Can you make sure your medics are ready, just in case? Okay, I'll rely on you.

MIKE JONES – CONTROL
16:19:50   Sir, if this is a hoax, then maybe it's a diversion for a theft attempt on those government missile figures. They're pretty important.

WILLIAMS – SECURITY CHIEF
Yeah, it could be. Let's tighten up security, just in case. Seal off the installation and double up guards on vital points.

MIKE JONES – CONTROL
16:20:10   Control to Gate 1.

GATE 1
Gate 1, receiving.

MIKE JONES – CONTROL
Gate 1, seal off your entrances. No one to go in or out until advised. Out.

GATE 1
Copy, Control. Ten-four, out.

MIKE JONES — CONTROL
Control to Gate 2.

GATE 2
Gate 2, receiving.

MIKE JONES — CONTROL
Gate 2, seal off your entrances. No one to go in or out until advised. Control out.

GATE 2
Copy, Control. Ten-four, out.

MIKE JONES — CONTROL
16:20:30 Control to Units 2, 3 and 4. Stand by for message.

He pauses briefly for the units to react, then continues.

16:20:33 Control to Units 2, 3 and 4. Go to planning office and reinforce guards. Be alert for an attack. Acknowledge in numerical order.

UNIT 2
Unit 2 to Control, en route. Out.

UNIT 3
Unit 3 to Control, en route. Out.

UNIT 4
Unit 4 to Control, en route, out.

MIKE JONES — CONTROL
16:20:50 Control to Units 2, 3 and 4. Acknowledged. Out.

WILLIAMS — SECURITY CHIEF
16:21:00 Okay, Mike, let's get Sergeant Connors to evacuate the area and make a quick search. We've still got fifteen minutes left. Get him some assistance down there.

MIKE JONES — CONTROL
16:21:05 Right, Sir. Mobile 12 has four guards uncommitted. (PAUSE) Control to Mobile 7.

MOBILE 7 — CONNORS
Mobile 7 receiving.

MIKE JONES — CONTROL
16:21:08 Mobile 7, evacuate personnel from computer room area. Conduct a bomb search and clear the area by 16:30. Assistance is en route.

MOBILE 7 — CONNORS
Copy, Control. Ten-four, out.

MIKE JONES — CONTROL
Control to Mobile 12.

MOBILE 12
Mobile 12 receiving.

MIKE JONES — CONTROL
16:21:15 Mobile 12, go to computer block and assist Sergeant Connors with a bomb search and evacuation. Over.

MOBILE 12
Copy, Control. Mobile 12 en route. Out.

MOBILE 7 — CONNORS
16:21:30 Mobile 7 to Control.

MIKE JONES — CONTROL
Go ahead, Mobile 7.

MOBILE 7 — CONNORS
16:21:33 Control, I've got personnel evacuated, and back behind cover. Assistance required in maintaining cordons. Over.

Dramatized Bomb Threat Security Response

16:21:37
    MIKE JONES — CONTROL
    Mobile 7, assistance is en route. Over.

    MOBILE 7 — CONNORS
    Control, be advised two guards are now searching the computer area with the building supervisor. Over.

    MIKE JONES — CONTROL
    Copy, Mobile 7. Use caution. Out.

16:22:00
    MOBILE 7 — CONNORS
    Mobile 7 to Control — emergency!

16:22:02
    MIKE JONES — CONTROL
    Go ahead, 7. All units stand by.

16:22:05
    MOBILE 7 — CONNORS
    Control, we've got a suspect device right in the computer room. Looks like a small gelignite and clock timer device. Over.

16:22:10
    MIKE JONES — CONTROL
    Copy, Mobile 7. Pull out your searchers and don't use radios nearby. I'll get the Bomb Squad there. Maintain your cordons. Control, out.

16:22:15
    WILLIAMS — SECURITY CHIEF
    Okay, Mike, I'm on the line now. Hello, Police Emergency. John Williams at InterComputer Security again. We've got a suspect explosive device in our computer room. Well, it's reported as a gelignite device with a clock time actuator.

16:22:17
    POLICE HQ. CONTROLLER
    Okay, we'll get the Bomb Squad down there in five minutes. Keep the area clear.

16:22:19
    POLICE HQ. DISPATCH
    Dispatch to Hotel Delta 1.

BOMB SQUAD COMMANDER BRADY
Hotel Delta receiving.

POLICE DISPATCH
16:22:22 Hotel Delta, go to InterComputer, 4520 West Avenue. See Security. Gelignite time device reported. Over.

BOMB SQUAD COMMANDER
Ten-four. On our way.

WILLIAMS — SECURITY CHIEF
16:22:25 Thanks. I'll arrange for an escort through our installation.

POLICE HQ. CONTROLLER
16:22:27 Fine. How long do we have before supposed detonation?

WILLIAMS — SECURITY CHIEF
We've got until 16:36. That's about... thirteen minutes left.

POLICE HQ. CONTROLLER
16:22:32 Okay. Well, if we can get in there fast, and it's a real device, we should just make it. Pray for easy traffic and keep your roads clear for the Squad.

WILLIAMS — SECURITY CHIEF
16:22:35 Right, I'll check on this end now. Call you later. Okay, Mike, the Bomb Squad is on the way. Should just make it in time, but we've got to get our traffic cleared. Any mobiles left free?

MIKE JONES — CONTROL
Mobiles 10 and 15 are uncommitted, Sir. Everyone else is on reinforced guard.

WILLIAMS — SECURITY CHIEF
16:22:45 All right, get Mobiles 10 and 15 down to Gate 1, to

clear traffic and escort the Bomb Squad through.

MIKE JONES — CONTROL
16:22:50 Control to Mobiles 10 and 15. Acknowledge in numerical order.

MOBILE 10
Mobile 10 receiving.

MOBILE 15
Mobile 15 receiving.

MIKE JONES — CONTROL
16:23:00 Mobiles 10 and 15, get up to Gate 1 and stand by to escort the Police Bomb Squad through. Get them through the traffic fast!

WILLIAMS — SECURITY CHIEF
16:23:10 I'm going to play safe, Mike. Those government computer tapes are worth a fortune to foreign agents. I'm going to run an ID check on the Bomb Squad before they pass into the installation. It's a very long shot, but we've got to be careful.

MIKE JONES — CONTROL
16:23:15 It shouldn't lose us more than ten seconds, Sir, and we know the device's position. Okay, I'll fix it up. (PAUSE) Control to Gate 1.

GATE 1
Gate 1 receiving. Go ahead, Control.

MIKE JONES — CONTROL
We have local Police Bomb Squad arriving in a couple of minutes. Get into position outside the gate and check their ID's — but make it fast. Mobiles 10 and 15 will be there to cover you and escort the Squad through. Over.

GATE 1
Copy, Control. Ten-four.

MIKE JONES — CONTROL
16:24:00   That's entry sewn up, Sir.

WILLIAMS — SECURITY CHIEF
Okay. Say, we'd better clear traffic at the incident site. We're running out of time and we've got to get the Squad through.

MIKE JONES — CONTROL
16:24:10   Control to Mobile 7.

MOBILE 7
Mobile 7 receiving.

MIKE JONES — CONTROL
16:24:12   Mobile 7, the Bomb Squad should be there in a few minutes. What do you copy on traffic?

MOBILE 7 — CONNORS
Mobile 7 to Control. We have traffic under control. Access is clear. Over.

MIKE JONES — CONTROL
Copy, Mobile 7. Control out.

GATE 1
Gate 1 to Control.

MIKE JONES — CONTROL
Control receiving, Gate 1.

GATE 1
16:24:30   Bomb Squad ID's check out okay. May they pass through? Over.

MIKE JONES — CONTROL
That's affirmative, Gate 1. Control out. (turning to Security Chief)

16:24:35 The Bomb Squad has passed through, Sir.

WILLIAMS — SECURITY CHIEF
We're cutting it real fine, Mike. Make sure Sergeant Connors is ready to brief the Squad Commander at the Incident Site.

MIKE JONES — CONTROL
16:24:43 Control to Mobile 7.

MOBILE 7 — CONNORS
Mobile 7 receiving.

MIKE JONES — CONTROL
16:24:52 Mobile 7, the Bomb Squad is en route to you through the installation. Brief their Commander. Control out.

MOBILE 7 — CONNORS
That's ten-four, Control. Out.

WILLIAMS — SECURITY CHIEF
16:25:00 Nothing to do now but bite your fingernails, Mike. How long to the supposed detonation?

MIKE JONES — CONTROL
16:25:04 We've got just about ten minutes, Sir, and about eight minutes after the Squad gets to the computer block.

WILLIAMS — SECURITY CHIEF
16:25:06 Let's hope it's a hoax, Mike.

MOBILE 7 — CONNORS
Mobile 7 to Control.

MIKE JONES — CONTROL
16:25:15 Go ahead, Mobile 7.

MOBILE 7 — CONNORS
Control, the Squad says the device is real. They're putting it into a bomb bucket right now. The timer is a phony, but they want to explode the gelignite outside the installation. Over.

MIKE JONES — CONTROL
16:25:35 Copy, Mobile 7. Have Mobile 10 escort the device out. I'll give you TX. Ten-five. (PAUSE) Go ahead, Mobile 7, you have TX.

MOBILE 7 — CONNORS
Mobile 7 to Mobile 10.

MOBILE 10
Mobile 10 receiving.

MOBILE 7 — CONNORS
Mobile 10, escort the squad from the installation. Use caution, they have a live explosive device. Over.

MOBILE 10
Ten-Four, Mobile 7. En route. Out.

MOBILE 7 — CONNORS
Mobile 7 to Control. TX completed.

MIKE JONES — CONTROL
Ten-four, Mobile 7. Control out.
 (turns to Williams)
16:26:00 Looks like we're okay, Sir. A phony mechanism but they're getting the explosive out of our installation area for detonation. That was a close one.

WILLIAMS — SECURITY CHIEF
16:26:05 Have the Bomb Squad declared the area safe so we can get the computers back to work?

Dramatized Bomb Threat Security Response

16:26:10	**MIKE JONES — CONTROL** Negative, Sir. Maybe they're still inside.
	MOBILE 7 — CONNORS (urgently) Mobile 7 to Control. Emergency.
16:26:15	**MIKE JONES — CONTROL** Go ahead, Mobile 7. Stand by all units.
16:26:17	**MOBILE 7 — CONNORS** Control, they've found another device in there. A parcel about two feet high, must have got past screening somehow. It's packed with explosive. Over.
	MIKE JONES — CONTROL Copy, Mobile 7. What is the ten-nine of the device? Over.
	MOBILE 7 — CONNORS Right in the main computer room, Sir, and next to the chemical stores. Over.
16:26:30	**MIKE JONES — CONTROL** Copy, Mobile 7. Evacuate all personnel back another 300 feet. Report as necessary. Over.
	MOBILE 7 — CONNORS Ten-four, Control. Out.
16:27:00	**MIKE JONES — CONTROL** They've found a real device, Sir. Nearly fooled us with that first phony. This sounds like at least a 30-pound device, right next to the chemical store. Our guys have pulled evacuees back another three hundred feet.
	WILLIAMS — SECURITY CHIEF Let's hope the Bomb Squad can render it safe in

132 BOMB SECURITY GUIDE

time. I guess Sergeant Connors briefed them on the value of those computer tapes and equipment.

MOBILE 7 — CONNORS
Mobile 7 to Control.

MIKE JONES — CONTROL
16:28:00   Go ahead, Mobile 7.

MOBILE 7 — CONNORS
Control, the Bomb Squad Commander is trying to render the device safe. His squad has pulled back to the cordon line. Over.

MIKE JONES — CONTROL
16:28:10   Copy, Mobile 7. Do you have contact with the Commander? Over.

MOBILE 7 — CONNORS
16:28:13   Negative, Control. He can't use radio near the explosive. Over.

MIKE JONES — CONTROL
Copy, Mobile 7. Advise if you hear anything. Control out.

MOBILE 12
16:28:20   Mobile 12 to Control.

MIKE JONES — CONTROL
Go ahead, Mobile 12.

MOBILE 12
16:28:28   Control, we've got four injured evacuees outside the computer block. They're in the danger area and people are panicking. Ambulance requested. Over.

MIKE JONES — CONTROL
16:28:35   Copy, Mobile 12. Keep things calm. Ambulance is

en route. If you have any more evacuation, tell people we have a gas leak being repaired. Control out.

MOBILE 12
Copy, Control. Ten-four.

MIKE JONES — CONTROL
Sir, some people panicked during the evacuation. They need an ambulance down there.

WILLIAMS — SECURITY CHIEF
16:28:40 Damn! Okay, I'll handle it, you keep the radio clear. How many injuries?

MIKE JONES — CONTROL
Four, Sir.

WILLIAMS — SECURITY CHIEF
(on telephone)
16:28:43 Hello, Sister Daniels. We've got four evacuation casualties at the computer block. Can you get an ambulance down there? Fine. Hey — tell them to lay off the sirens. I don't want any more panic.

MIKE JONES — CONTROL
Control to Mobile 7.

MOBILE 7 — CONNORS
Mobile 7 receiving.

MIKE JONES — CONTROL
16:29:00 Mobile 7, an ambulance is en route to your casualties. Hold down any panic. Anything to report? Over.

MOBILE 7 — CONNORS
Negative, Control.

MIKE JONES — CONTROL
16:29:08 Control out. Nothing yet, Sir. The Commander can

only have a few seconds left. He should pull out now.

WILLIAMS — SECURITY CHIEF
16:29:12 Let's hope he's got it rendered safe. If our chemical store goes up there will be toxic fumes everywhere. Let's have a mobile stand by to escort the Commander out if he succeeds. Then I'll place the area disaster plan on alert.

MIKE JONES — CONTROL
Right, Sir. Mobile 15 is at the incident site and uncommitted. (PAUSE) Control to Mobile 7.

MOBILE 7 — CONNORS
Go ahead, Control.

MIKE JONES — CONTROL
16:29:25 Mobile 7, have Mobile 15 put his vehicle on stand-by to escort the Commander out.

MOBILE 7 — CONNORS
Copy, Control. Be advised the Commander has rendered safe the device's main charge but its initiator is still live. He's got to detonate it in the open within three minutes. Over.

MIKE JONES — CONTROL
16:29:34 Copy, Mobile 7. Get Mobile 15 on the road *fast*, and get them clear.

MOBILE 15
16:29:37 Mobile 15 en route.

MIKE JONES — CONTROL
Control to Gate 1.

GATE 1
16:29:45 Gate 1 receiving.

MIKE JONES — CONTROL
16:29:49 Gate 1, Mobile 15 will be escorting the Bomb Squad Commander out with a live explosive device. Get traffic clear and get them through fast.

GATE 1
16:29:53 Ten-four, Control. Traffic is clear. Out.

MIKE JONES — CONTROL
16:30:00 If we can keep the next couple of minutes running smoothly we'll be clear, Sir.

WILLIAMS — SECURITY CHIEF
16:30:04 Right. How's traffic?

MIKE JONES — CONTROL
16:30:08 Reported clear, Sir. We've got about one minute left for him to clear the installation and detonate the device.

GATE 1
Gate 1 to Control.

MIKE JONES — CONTROL
16:30:25 Go ahead, Gate 1.

GATE 1
Bomb transporter and Mobile 15 passed through, Control. All secure. Over.

MIKE JONES — CONTROL
16:30:28 Copy, Gate 1. Control out.

MOBILE 15
Mobile 15 to Control.

MIKE JONES — CONTROL
Go ahead, Mobile 15.

MOBILE 15
16:30:40   Control, the device has been safely detonated. Over.

MIKE JONES — CONTROL
16:30:43   Copy, Mobile 15. Return to station. Control out.
(turning to Williams)
Device detonated safely. That was a close one, Sir.

WILLIAMS — SECURITY CHIEF
16:30:47   You're right, we only had 45 seconds left.

There is a moment of silence in which the two men regard each other. There is no need to express their relief.

Okay, let's get the place back to normal.

MIKE JONES — CONTROL
Control to Mobile 7.

MOBILE 7 — CONNORS
Mobile 7 receiving.

MIKE JONES — CONTROL
Situation report, Mobile 7?

MOBILE 7 — CONNORS
Incident area declared safe after final search by Bomb Squad. Ready for reoccupation. Casualties removed. Over.

MIKE JONES — CONTROL
Copy, Mobile 7. Go ahead and reoccupy. Reassure people that the area is safe and try to prevent false rumors. Over.

MOBILE 7 — CONNORS
Copy, Control. Ten-four.

16:33:00

MIKE JONES – CONTROL

Ten-four. Control out. (PAUSE) Control to all units. Return to normal stations. Well done, everybody.

WILLIAMS – SECURITY CHIEF

Let's have some coffee, Mike, and stand down emergency services.

Appendix C

OPERATION OF X-RAY INSPECTION EQUIPMENT

In key installations, and at air or sea ports, luggage and cargo must be inspected to prevent the infiltration of weapons, explosives or other hazardous devices. X-ray inspection systems are used to reduce delays in luggage inspection, through the elimination of time-consuming hand searching.

It is essential that search operatives be alert and watchful. Searching hundreds of cases every hour can be very monotonous, but the one case you overlook could contain a bomb which kills a hundred people. If you become tired or bored, take a few minutes break.

The primary rule of searching (included in the instructions of the Federal Aviation Authority) is to allow past only those items which you can *positively* identify. Any object which you cannot positively identify must be removed for hand searching.

Among the keys to effective X-ray screening are the following:

1. Watch for the outline of explosive devices: timer clocks or watches, wire, detonators, batteries, switches and the blurred outline of explosives.
2. Watch for the outline of incendiaries: bottles, cans, fuses, etc.
3. Watch for the outline of firearms, knives and razor blades. Check that these are not hidden behind other harmless metal objects.

4. Check cans for tear gas. Tear gas canisters have no seam at the side or top, whereas cans containing hair sprays and deodorants do.
5. If you spot an object which you cannot positively identify, remove it for hand searching.
6. If you spot a suspect hazardous device or weapon, call for the law enforcement or security officer on duty.
7. *Stay alert*—other people are relying on you for their lives; don't let them down.

Some of the more common items that will be encountered in using the inspection system are listed in the following table. A number of suspicious objects are also given. The inspection system operator should become familiar with as many objects as possible to eliminate the need for unnecessary visual inspection.

Normal Items		Suspicious Items
• Aerosol cans	• Toys	• Batteries
• Clothes hangers	• Bottles	• Wiring
• Hair brush	• Radios	• Electronic devices
• Razor blades	• Shavers	• Barostats
• Key	• Glasses	• Motors
• Clothing with zippers	• Soap	• Bullets
	• Dentures	• Hypodermics
• Clocks	• Pipes	• Vials
• Shoes	• Cosmetics	• Medicines
• Pens	• Flashlight	• Cylindrical objects
• Coins	• Buckles	• Springs
• Scissors	• Books	• Dense objects - any shape
• Cameras	• Tape recorders	
• Cigarette lighters	• Jewelry	• Objects in luggage lining

Figure 14. X-Ray Detectable Luggage Items

Appendix D
BOMB THREATS — POLICE RESPONSE PROCEDURE

One of the main purposes of bomb threats is to waste an organization's time and money. Police response to a bomb threat may require the deployment of officers for searching, evacuation and crowd control, traffic control, suspect device inspection, removal and neutralization, plus follow-up investigations. An inexpensive telephone call can cost a police department thousands of dollars in time, and distracts manpower from other patrol duties. These factors make a fast and efficient police bomb threat response essential in the interests of safety and economy.

COUNTER BOMB THREAT PROGRAM

There are five phases to a police counter bomb threat program. As with many other law enforcement and security tasks, prevention is better than cure. Preventive security is the first phase of our program. The second phase is intelligence; the third is planning. Fourth is threat response, followed by analysis.

Preventive Security

Preventive bomb security should be incorporated into a police department's Crime Prevention program. The Crime Prevention division should recommend preventive security programs for those organizations in their region which have not already initiated them. Recommendations should include advice on perimeter security and entry control systems, internal security protection, plus specific planning to deal with telephoned bomb threats. Crime prevention outreach programs are cost-effective, because secure organizations are less likely to receive bomb threats or to require police assistance in dealing with them. Considerable savings in both direct costs and manpower deployment have been effected where organizations have adopted effective bomb threat security systems, and police departments should encourage their implementation.

Intelligence

The intelligence phase is divided into two sections: active and passive. The active phase should be accomplished by the department's Criminal Intelligence division. This division should gather intelligence on potential bombings and threats through overt and covert methods. When operationally effective, this intelligence should be made available to endangered organizations, directly or indirectly, so that they can implement or improve their security systems.

The passive phase of intelligence should be accomplished by the department's Bomb Squad or Anti-Terrorist Squad. This division should monitor reports from the Criminal Intelligence division. Either the Bomb Squad or the Crime Prevention division should recommend security systems to endangered organizations. Secondly, the Bomb Squad should identify potential bomb threat targets in their region and request that the Crime Prevention division recommend additional security for those organizations. Identification of potential targets will be made by size of organization, nature of business, type of event, political or racial factors, life and structural risks, etc.

Planning

The planning phase incorporates many factors from the intelligence and preventive security phases. It also involves an interaction between organizations (or their contract security forces) and the police. The private and public security functions should interact so as to present a unified and cost-effective bomb threat response. After the initial advice of the Crime Prevention division, the joint public/private security approach should become the responsibility of the police department's Patrol Division and Bomb Squad.

Threat Response

The Patrol Division will be responsible for bomb threat response and basic investigation. The Bomb Squad will be responsible for the inspection, removal, destruction or neutralization of suspicious and hazardous objects.

Both the Patrol Division and the Bomb Squad should be prepared to develop a coordinated plan for bomb threat responses with private organizations or their contract guard services. This will involve the development of a unified reporting procedure for emergencies and the establishment of divisions of responsibility for searching, crowd and traffic control, and evacuation control. Additionally, Patrol Divisions and Bomb Squads can be provided with detailed plans of organizations and information on special risk areas, such as fuel or chemical storage (which may require the preparation of a Disaster Plan by relevant emergency forces). In a high-risk organization, or where convenient to both parties for training or evaluation programs, a joint bomb threat response exercise may be organized to practice and test plans.

Interaction and cooperation should involve other agencies in addition to the private organization and police force. It should include other emergency services, such as the fire department, hospital, civil defense, military forces, etc. These organizations should coordinate the preparation of emergency plans, and whenever possible mount joint practice or evaluation exercises.

Analysis

The final phase is analysis, where the police, preferably in conjunction with organization security directors and chiefs of other involved emergency services, meet to discuss, evaluate and criticize bomb threat response incidents or exercises. This will assist in the analysis of police response performance, and also in updating/ improving the organization's preventive and emergency security programs.

RESPONSE PROCEDURE — "I.E.A.A."

There are four phases to a police bomb threat response: information, evaluation, action, and analysis.

A bomb threat may be reported directly to an organization, or to the police department. Either way, foot or vehicle patrols will be informed of the threat by the radio dispatcher. The dispatch message will probably be brief and unspecific—for example, "Bomb threat at . . . , device reported, see the supervisor." However, it may be more specific and informative, such as, "Bomb threat at . . . , device reported in computer room, supposed to detonate at 16.20, see the security guard."

Upon receiving and logging the dispatch message, the patrol unit should respond, give an ETA to the incident site, and proceed by Code 3 if necessary (red light and siren). As with all other emergency situations, the emergency should not be placed before reasonable driver safety; remember, it is better to take an extra minute to get there than not to get there at all.

If possible, the responding officers, especially the Recorder, should watch out for persons acting suspiciously near telephone booths along the incident approach route, or persons leaving the area suspiciously by foot or vehicle. At this stage, pursuit of suspicious persons should not be attempted, as the primary response role is to prevent injury and damage from a possible bombing. However, descriptions of suspicious persons and nearby witnesses should be noted.

Upon arrival at the incident site, the patrol vehicle should be parked far enough away from the bomb threat site to avoid possible blast damage, and also out of the access route of other responding emergency vehicles.

It must be remembered that a bomb threat or even a small fire or explosion may be only a diversion for a robbery, or a lure to draw responding personnel into sniper fire or the kill zone of explosive devices.

Therefore, it may be advisable to park the patrol vehicle further away from the incident site than might be expected, and preferably to park in or near cover. Before leaving the patrol vehicle, officers should check for potential ambushes and snipers. Places of cover and concealment should be checked, such as rooftops, windows, bushes and trees, alley corners and stairways, etc. Similarly, the officers should keep a watchful eye on these locations when approaching the incident site.

An additional precaution is to scrutinize the incident site's access path for trip wires, suspicious objects or mines, which may have been planted specifically to injure responding police officers. Such devices may be activated by the movement of police officers or their vehicles, or by direct/remote control of a distant observer.

Information

When the responding officers have entered the incident site, observing these basic precautions, they should quickly contact:
1) the acting supervisor/security chief, and
2) the person who took the threat call.

At this point the most important information which the officers must discover is:
1) What is the time of the supposed explosion?
2) What is the location of the supposed bomb(s)?

The next actions taken by the officers will depend upon the potential target risk, and the answers to the two questions above. If the information is that a bomb has been planted near a fuel store, and is detonating in less than five minutes, then they will probably decide to evacuate the site immediately. If, however, the threat is less dangerous, such as a device supposed to detonate in twenty minutes, and the organization could suffer little loss of life, then an investigation and search would probably be organized.

In addition to the two questions above, the police officers should ascertain whether any suspicious object has been reported by the organization's security guards or staff. If so, then the bomb threat becomes more dangerous and may require an immediate evacuation and the assistance of the Bomb Squad.

Evaluation

Assuming that the initial information suggests a safe amount of time, the threat information should be evaluated. This evaluation phase requires that the person who received the threat call should be questioned. After ascertaining the supposed time and place of the explosion, the police should seek the following information:

- Was the call from a phone booth?
- What was the telephone operator's impression of the caller's mental state?
- Were there any background noises?
- What were the exact words used by the caller?

To evaluate this information, the following points provide the basis for planning a response:

1. Do the time and location of the supposed device indicate sufficient risk to suggest an immediate evacuation?
2. Is immediate evacuation possible or safe?
3. Is there reason to consider the threat real (code words used in threat call, known terrorist activity, intelligence reports, etc.)?
4. Is there reason to consider the threat a hoax (a recently dismissed employee, recently dissatisfied or ejected customer, drunken voice, children laughing, etc.)?
5. Is there time to investigate/search before the supposed explosion?

It should be remembered that perhaps only 2% of reported bomb threats are real, and that their major effect is loss of time, production and money. Because of this, evacuation is an undesirable response to many bomb threats; in the interests of economy it should be avoided whenever possible. However, where the five questions above suggest a real threat, financial loss should never be placed above human life. If a threat sounds real, or there is too little time to investigate or search, then evacuation is necessary.

POSSIBLE RESPONSES

Depending on the results of the Information and Evaluation phases, there are five possible responses:

1. Handing over incident control to the Bomb Squad.
2. Full evacuation.

3. Localized evacuation.
4. Staff search (police-supervised).
5. Police search

Control by Bomb Squad

Under exceptional circumstances, the incident may be handed directly to the Bomb Squad. This will usually occur when the area is suspected to be booby-trapped, when special search equipment is required, or when a suspicious object has already been discovered. (If a suspicious object has already been located by the organization's staff by the time police officers arrive, the area surrounding the object should be evacuated, cordon lines established, and the other *safely distant* areas checked for further devices.)

Full Evacuation

If a suspicious object of sufficient size is located, or there is insufficient time to search, or vulnerable points are at risk, then a full evacuation may be necessary.

Evacuation must be accomplished quietly and without panic. If possible, a building's occupants should be told by their supervisors that there is a power leak or gas leak which must be repaired. This, or some other equally plausible story, will help to avoid confusion and panic.

Whenever possible, the emergency evacuation routes and assembly routes should be searched for suspicious objects or car bombs, etc. A devious bomber could obtain maximum casualties by planting a device in a corridor or assembly area just prior to an evacuation.

It is essential to avoid panic and confusion when organizing an evacuation; therefore, the organization's supervisors and security guards should be requested to assist police officers in the maintenance of orderly evacuation. Evacuation should strictly follow an established and proven set of safety rules, provided below.

All personnel should leave through main or fire exits, in a quiet and orderly manner. A walking pace should be used. Running will only cause panic and injuries.

Personnel should assemble in the pre-arranged safe areas after evacuation. Supervisors will then hold a roll call to ensure that all personnel and visitors are accounted for. Any person not present will be immediately reported to the police or security.

As they leave, supervisors should try to disconnect any electrical apparatus, such as calculators, fans, heater, typewriters, etc.

Building custodians or supervisors should switch off major plant equipment, such as air conditioning, heating, extra lighting, etc. To prevent time loss during an evacuation, this may be accomplished later by remote switching from a master control.

Elevators or lifts should *not* be used during evacuation.

As personnel evacuate, they should be instructed to take their personal belongings with them, provided this can be accomplished quickly.

As personnel evacuate, they should be instructed to open all possible doors, windows, cupboards, filing cabinets, etc. Supervisors, who should leave the area or building last, should try to open as many doors and windows as possible.

Evacuated personnel should not be allowed to congregate and hinder the access of emergency vehicles. If this risk becomes likely, then assistance should be requested for crowd and traffic control.

Evacuated personnel should be positioned at least 300 feet (100 meters) away from suspect areas, and if possible placed behind blast-proof cover.

Localized Evacuation

A localized evacuation may be used around the specific building or area mentioned in a threat call, or around a suspicious object. The usual evacuation distance for a small object is a mimimum of 300 feet (100 meters). When a device is in a multi-story building, then the floors above and below the reported object should also be evacuated.

Staff Search

A staff search is the minimum response to a threat call; however, it is very efficient. A building's staff will quickly detect any strange or out-of-place objects whereas everything will be strange to an officer walking into the building for the first time.

When a staff search is recommended by the responding police officers, the staff must be warned by the officers or their supervisors not to touch any suspicious objects. A staff search is usually restricted to the employees' immediate work areas, with other areas checked by supervisors and/or the police officers.

Wherever possible police should supervise and assist in a staff search, in the interests of safety—and to save time. Police officers should search the potential bomb planting areas, such as reception foyers, lavatories, trash cans, locker rooms, vulnerable points, etc.

Police Search

A police search should be used when:
a. building staff are unavailable or unwilling to search;
b. there is a high risk element;
c. checking a building before allowing re-occupation.

The police search must be made with full attention placed upon safety. There are, in fact, a set of safety rules for searching which must be strictly followed. They are as follows:

Search Safety Rules

1. Never use more searchers than necessary.
2. Use a maximum of two searchers for each room, or for an area up to 250 square feet.
3. Use searchers in alternate rooms or areas.
4. Never assume that only one device has been planted; even if one device is found, continue searching until the entire area has been cleared.
5. Clearly mark and report areas searched and cleared.
6. Clearly mark and report areas found hazardous.
7. Do not allow searchers to work for more than ten minutes without a few minutes' break.

If sufficient police personnel or other personnel are involved with searching, then a search incident control point should be established for liaison, direction and communication. The incident control point should be set up in a convenient safe area, preferably where a trunk phone line exists.

Searching should be accomplished as follows:

1. Areas reported in the threat call.
2. Exterior of building (trash cans, foliage, gardens, window ledges, drain/air inlet ducts, etc.)
3. Interior of building (primary search in areas readily available to the public, such as reception areas, lavatories, locker rooms, recreation facilities/canteen, etc.).

An interior search should be accomplished methodically and carefully, using one of the following methods:

Search Technique 1—Split the room into two halves. Search each half separately, or if you have assistance, search one half each, working back to back. Then search each half in three sections, floor to waist level, waist to eye level, and eye to ceiling level.

Search Technique 2—Search the room in the following sequence: floors; furniture; walls; fittings/decorations; permanent fittings.

General search procedures call for moving slowly and carefully, so as to avoid booby traps. As you enter a room, pause and listen. You may be able to hear a ticking clockwork device, or a spluttering fuse. (This is the reason for switching off all machinery/electrical equipment during bomb threat evacuations.)

Suspicious Object/Hazardous Device Procedures

If a suspicious object or hazardous device is found by search, or reported upon the officer's arrival at the scene, then the Bomb Squad should be called (from safe radio range), and responsibility handed over to them. There is a great deal which responding officers can do to assist the Bomb Squad's task. First they should evacuate endangered areas to at least 300 feet (100 meters), and establish cordons to keep traffic access clear. Second, they should draw a sketch map of the location of the device(s), its precise description, size, etc. They should also show on this map the proven safe route to the object. Then an Incident Control point should be set up (if this has not already been done for the search operation) for communications, control, etc. The Incident Control point should be established in a safe position, preferably one which allows access for emergency vehicles. The Incident Control point will be commanded by the Senior Officer responding, or, where appropriate, by the Bomb Squad commander.

POLICE RESPONSE PROCEDURE — SUMMARY

1. Log Dispatch Message.
2. Watch out for suspicious persons near phone booths, or leaving the area suspiciously. Note descriptions of suspicious persons and witnesses, but do not pursue at this time.
3. Park a safe distance from the Incident Site.
4. Do not park where you will restrict access of other emergency vehicles.
5. Try to park behind or near cover.
6. Be alert for snipers and ambush.
7. Be alert for mines, trip wires, booby traps and suspicious objects on your approach route and path to the Incident Site.
8. Contact the organization's supervisors or the person who took the threat call.
9. Find out the supposed time and place of explosion.
10. Evaluate the degree of threat risk, based on information provided by the person who took the call.
11. Decide on a) full evacuation
    b) localized evacuation
    c) staff search
    d) police search
12. Set up an Incident/Search Control Point.
13. Is the Bomb Squad necessary?
14. Control evacuation to avoid panic or confusion.
15. Position evacuees behind solid cover, a minimum of 300 feet (100 meters) away from hazardous areas.
16. Strictly observe Search Safety Rules.
17. Do not touch any suspicious object/device. If devices are found or suspected, report them to the Bomb Squad, from a safe radio distance.
18. Do not allow building re-occupation until the entire area has been searched and found clear. Do not stop searching after the discovery of one or two devices or suspicious objects.
19. Maintain communications with your Control/H.Q. in the event of support requirements, disaster/emergency service requirement.
20. The job of responding officers is immediate decision-making, and the safeguarding of life through evacuation or fast search. Do not lose your control of the situation by attempting other specialized tasks; request support assistance where appropriate.

INDEX

Activation methods, 40-42, 93, 105-110, 113-114
    anti-personnel handling, 41-42, 104-106, 109-110
    of booby traps, 41-42, 104-110, 113-114
    chemical, 40, 41
    electrical, 41, 42
    of letter bombs, 93
    mechanical, 40-41
    time delay, 40-41, 105, 109
Alarm systems, for perimeter security, 14, 82
Ambulance services, 57
Anti-personnel handling devices, 89
    (See also Booby traps; Letter bombs)
Arson, 2, 4, 6, 9, 81
    (See also Incendiary devices)

Black powder, 34
Blasting caps, 42, 107
Blast suppression blankets, 69, 100
Bomb bunkers, 100
Bomb Squad, 62, 65, 67, 74, 142, 143, 147, 150
    (See also Law enforcement response to bomb threats)

Booby traps, 41-45, 67, 74, 89-90, 103-116
    components of, 106-110
    diversion methods, 111-112
    external, 111-112
    internal, 112-115
    in machinery, 113
    motives of perpetrators, 103-104
    recognition clues, 42, 115-116
    targets of, 104
    types of, 104-106
    in vehicles, 22-23, 70, 74
Booster charge, 33-34

Card-operated entry systems, 20, 23
Chemical activation, 40, 41
Chemical igniters, 108
Cigarettes, as improvised fuses, 108, 109
Closed circuit television, 17-18, 22, 23, 28
    in internal security, 28
    in perimeter security, 17-18
    in vehicle control, 22, 23
Corridors, as bomb planting areas, 25, 65
Crowd control, 62

154   INDEX

Delivery supervision, 22, 23, 70, 91, 92-93
   (See also Mail and parcel monitoring)
Detectors, explosives, 65, 91-93
   of letter bombs, 93, 96
   (See also Fluoroscopes; Gas Chromotography detectors; Metal detectors; Vapor detectors; X-ray inspection devices)
Detonators, 33, 92, 93
Disaster plans, 57
Dogs, vapor-detecting, 19, 65
Dynamite, 39

Electrical activation, 41, 107
Electrical blasting caps, 42, 107
Elevators, 64, 148
Emergency contact list, 52
Entry control, 11, 18-23, 84-85
   automatic entry systems, 20, 23
   deliveries, 22, 23, 70, 91, 92-93
   entry log, 20-22
   of staff, 19-20, 23
   of vehicles, 22-23
   of visitors, 20
   (See also Perimeter security)
Evacuation, 10, 56-64, 70, 82, 83, 86, 100, 147-148
   clearing routes and assembly areas, 60
   full, 56, 59, 147-148
   injuries during, 10, 59-60
   and letter bombs, 100
   localized, 56, 59, 148
   methods and instructions, 62, 64
   police assistance with, 147-148
   recommended distances, 62-63
   warnings, 60-61
Executives, as targets of letter bombs, 90, 91
Explosions
   minimizing damage of, 69-70
   risks of, 10
Explosive devices, improvised, 8, 34, 103-116
   components of, 106-110
   diversion methods, 111-112
   external, 111-112
   internal, 112-115
   motives of perpetrators, 103-104
   recognition clues, 115-116
   sources of materials, 8, 34, 106-107
   targets of, 104
   types of, 104-106
Explosives
   clues to recognition, 69, 93, 96-97, 99
   types of, 34, 39, 107, 108-109
      in booby traps and improvised devices, 107, 108-109
      high, 39
      low, 34

Federal Bureau of Investigation, Uniform Crime Reports, 1, 3, 5, 47
Fences
   for perimeter security, 14
   in vehicle control, 23
Fire alarm, for evacuation warning, 61-62
Firing train, 33-34
Fittings, internal, as hiding places for bombs, 25, 113-114
Fluoroscopes, as letter bomb detectors, 91-92
Foams, anti-explosive force, 69
Frictional igniters, 108
Fuses, 41, 92, 93, 96, 108

Gas chromotography detectors, 19
Gasoline, as explosive or incendiary, 23, 40
Gelignite, 91, 107
Government agencies, assistance in bomb threat response, 83-84
Grilles, to protect windows and other openings, 17, 25
Guards, 14, 18, 20, 22, 23, 24, 25, 55, 78, 82-83

contract services, 82-83
  in entry control, 18, 20, 24
  in internal security, 25, 28
  in monitoring deliveries, 22
  in perimeter security, 14
  in response to bomb threats, 55, 78
  searches by, 24
  in vehicle control, 23
Guerrilla warfare, 1, 2, 6, 103
Gunpowder, 34, 93

High explosives, 39

Identification cards, 20
Identification checks, of authorities responding to bomb threat, 59
Igniters, 107-108
Incendiary devices, 9, 17, 39-40, 41, 81, 93
  activation methods, 41
  in letter bombs, 93
  in sabotage, 9
  (See also Arson)
Incident Commander, in response to bomb threat, 55, 56, 62, 76-80
Injuries
  in bombings, statistics on, 1-3
  during evacuation, 10, 59-60
Intelligence gathering, police role in, 8, 142
Internal security, 24-25, 28
  and design, 25
  openings, 25
  visibility, 28
  vulnerable areas, 25, 28-29

La Guardia Airport bombing, 2, 3
Law enforcement response to bomb threats, 8, 57, 59, 62, 67, 74, 82, 100, 141-151
  assistance in developing corporate policy, 83-84
  and evacuation, 147-148
  and letter bombs, 100
  and search, 57, 59, 65, 148-150
  summary of, 151

Letter bombs, 29, 89-101
  clues to recognition, 93, 96-97, 99
  detection devices for, 91-93
  security response to, 99-102
  target risk evaluation, 89-90
  training personnel in handling, 90-91, 101-102
  (See also Delivery supervision; Mail and parcel monitoring)
Lighting, 14, 23
  for perimeter security, 14
  and vehicle control, 23
Locks, pushbutton combination, 20
Low explosives, 34
Luggage, inspection of, 19, 20, 29, 93, 139-140

Machinery, booby traps in, 9, 113
Mail and parcel monitoring, 18-19, 22, 90-102
  (See also Delivery supervision; Letter bombs)
Management's role in bomb threat response, 29, 56-57, 81-86
  in smaller organizations, 81-86
Mechanical activation, 40-41
Metal detectors, 19, 22, 24, 91-92
  in entry control, 19, 24
  in mail and parcel monitoring, 22, 91-92
Military assistance in developing corporate policy, 83
Military components of booby traps, 107
Mines
  (See Booby traps; Explosive devices, improvised)
Molotov cocktails, 117
Motives of bomber/hoaxer, 4-5, 7, 103-104

Nitrocellulose, 34
Nitroglycerin, 39

Parking control, 19, 22-23, 29, 70

for staff, 23
for visitors, 23
Patrols, 14, 23, 28, 85
   in internal security, 28
   in perimeter security, 14
   in vehicle control, 23
Perimeter security, 13-23, 82
   closed circuit television for, 17-18
   for entry control, 18-23
   outer barriers, 14
   and vehicle control, 22-23
   windows and other openings, 17
Pipe bombs, 34
Plastic explosives, 39, 107
Propellants, as igniters, 108
Property damage in bombings, statistics on, 1-3
Public address system, for evacuation warning, 61
Pyrotechnic blasting caps, 42, 107
Pyrotechnic fuses, 108

Radios, activation of explosive devices by, 42, 106, 110, 112
Reception areas, as bomb planting areas, 25, 65, 85
Receptionists, in entry control, 18, 20, 22, 24, 85
Rest rooms, as bomb planting areas, 25, 65, 85
Robbery, bomb threat as diversion for, 57

Sabotage, 6, 9, 90, 104, 109, 113
   and booby traps, 104, 109, 113
   forms of, 9
   psychological, 9, 90
Safety precautions in bomb threat response, 45, 65-67, 99-102, 149
   and letter bombs, 99-102
   in search, 65-67, 149
Sandbags, 69, 102
Searches
   in entry control, 19, 22, 24, 29
      acceptability of, 24
      hand searches, 19, 24
   legal limitations on, 24
   of vehicles, 19, 22, 23
   of vulnerable areas, 29
   for suspected explosive device, 56, 59, 61-62, 77, 79, 83, 86, 148-150
      clues to recognizing explosives, 69
      by employees, 56, 59, 65, 67, 148-149
      by police, 56, 59, 65, 149
      report form, 68
      safety precautions for, 65-67, 149
      by security personnel, 56, 59, 65, 67
      techniques, 67, 69, 150
   of vehicles, 70, 74
Security consultant, 82-83
Security response to bomb threats, 8, 11-12, 29-31, 55-80, 99-102, 117-137
   dramatization of, 117-137
   evacuation, 59-64
   follow-up investigation, 74
   letter bombs, 99-102
   management's role in, 29, 56-57, 81-86
   search, 64-74
   security director's role in, 55, 57, 75-80
   in smaller organizations, 81-86
   staff deployment, 29-31, 84-86
   summary of, by personnel, 74-80
   target risk evaluation, 9-11
Statistics on bombings, 1-3, 5, 47

Tape recorder, to record bomb threat calls, 49
Targets of bombings
   of booby traps, 104
   high-risk, 3-4
   of letter bombs, 89-90
   risk evaluation of, 9-11
   types of, 3-4, 5, 8

Telephone company, tracing of calls, 48-49
Telephoned bomb threats, 10, 11, 30, 47-52, 71-73, 75-80, 81-82, 85-86
  effects of, 10, 11, 81-82
  emergency contact list, 52
  operator's response, 75
  procedures report, 71-73
  report for, 51
  staff training in response, 30
Telephone operator, in response to bomb threats, 30, 48-52, 71, 75, 145
Terrorism, 1-2, 4-8, 11, 28-29, 39, 89, 103, 107
  and booby traps, 103, 107
  and incendiary devices, 39
  and letter bombs, 89
Thermal igniters, 107-108
Time delay activation, 40-41
Training in bomb threat response, 10-11, 24, 29-31, 48-53, 90-91, 101-102
  in entry control, 24
  in handling letter bombs, 90-91, 101-102
  for security personnel, 10-11, 24, 29-31, 82
  for telephone operator, 48-53
Trinitrotoluene (TNT), 39, 107

Unexploded devices, 45, 112
U.S. Capitol bombing, 2

Vapor detectors, 22, 23, 91, 93
  in mail and parcel monitoring, 22, 91, 93
  in vehicle control, 23
Vehicles
  activation of explosive devices by movement of, 106
  control and parking, 19, 22-23, 29, 70
  delivery, search of, 22, 23
  explosive devices in, 22-23, 70

  police, in responding to bomb threats, 144-145
  searches of, 19, 22, 23, 70, 74, 93
  of staff members, 23
  of visitors, 23
Vulnerable areas, bomb planting in, 25, 27-29

Walls, for perimeter security, 14
Windows and other openings, and perimeter security, 17

X-ray inspection devices, 19, 22, 91, 92-93, 139-140
  detectable luggage items, 140
  in entry control, 19
  in mail and parcel monitoring, 22, 91, 92-93

Other Security World Books of Interest. . . .

HOTEL & MOTEL SECURITY MANAGEMENT
By Walter J. Buzby II and David Paine (248 pp.)
In-depth coverage of hotel-motel security hazards such as theft, holdup, fraud, bomb threats, fire, disaster, special events, etc., plus U.S. and Canadian laws affecting hotels. For hotel and motel management, students of hotel administration, security personnel.

HOSPITAL SECURITY
By Russell L. Colling (368 pp.)
Complete protection of people and property in health care facilities. A detailed, practical program for establishing or refining security systems to deal with theft, assault, kidnapping, fire, disaster, strikes, more.

INTERNAL THEFT: INVESTIGATION & CONTROL
An Anthology (276 pp.)
Two dozen top security professionals analyze Why Employees Steal, Executive Dishonesty, Embezzlement, Undercover Investigation, Interrogations, Confessions, Polygraphing, Pre-Employment Screening, etc. 25 chapters.

CONFIDENTIAL INFORMATION SOURCES: PUBLIC & PRIVATE
By John M. Carroll (352 pp.)
A unique, behind-the-scenes guide to the confidential personal information in public and private records. Reveals what is on file, how it is gathered, who has access, how to identify unknown persons.

INTRODUCTION TO SECURITY
By Gion Green and Raymond C. Farber (338 pp.)
A comprehensive introduction to the history, nature and scope of security in modern society, with basic principles of physical security, internal loss prevention, defensive systems, fire prevention and safety, etc.

SUCCESSFUL RETAIL SECURITY
An Anthology (303 pp.)
Employee theft, shoplifting, robbery, burglary, shortages, special fire problems, insurance recovery – 25 top security and insurance professionals pool their expertise in retail loss prevention.

OFFICE & OFFICE BUILDING SECURITY
By Ed San Luis (295 pp.)
The first book devoted exclusively to the staggering security problems, both internal and external, facing today's offices, high-rise office buildings and personnel, with practical solutions.

ALARM SYSTEMS & THEFT PREVENTION
By Thad L. Weber (385 pp.)
A highly readable guide to the entire gamut of alarm systems and problems. Strengths and weaknesses, how alarms are attacked by criminals, and techniques required to defeat attacks are offered in laymen's terms.

AIRPORT, AIRCRAFT & AIRLINE SECURITY
By Kenneth C. Moore (356 pp.)
Definitive study of every aspect of air traffic security today, from the hijacking threat to predeparture screening and baggage handling; airport physical protection; general aviation problems; air carrier security including ticket theft, credit card fraud, internal theft and investigation management; and air freight security procedures.

In addition to its hard cover books on security subjects, Security World Publishing Company publishes *Security World* and *Security Distributing & Marketing (SDM)* magazines; produces booklets, manuals and audio tape cassettes on security; and sponsors the International Security Conference. Books and other materials are available from Security World Publishing Co., Inc., 2639 So. La Cienega Blvd., Los Angeles, California 90034.